SEAN DORMAN, as a boy o'
Irish public school, was award
and the English master regular
Form. He became editor of t
winner of an essay competiti
Great Britain and Ireland. After graduating at Oxford,
as a freelance journalist in London, having articles published in
some twenty British and Irish periodicals. He also ghosted half
a dozen non-fiction books for a publisher. He had a burlesque
of Chekhov staged by the Dublin Gate Theatre, a radio play
broadcast by Radio Eireann, and he published a few short stories,
one being broadcast by the BBC. For five and a half years
in Dublin he published a literary, theatre and art magazine,
Commentary. In England, from 1957 until a few years ago, he
ran the Sean Dorman Manuscript Society for mutual criticism, a
Society still run by others under the same name and listed in *The
Writers' and Artists' Year Book*. His magazine *Writing*, also
listed during its career in 'The Year Book', was founded in 1959,
and sold after twenty-six years. While his family was growing
up, he had to seek a more regular income, and taught secondary
school French and junior German for some twenty-five years. To
make up for lost time, between 1983 and 1993 he wrote and
published, under the imprint of his Raffeen Press, eleven books.
They embraced novels, autobiography, essays, a three-act play,
theatre criticism, short-stories and a solitary poem. Many of the
books were re-writes or extensions, now allowed to go out of
print, so the present tally is six. Five of these have been included
in his three-volume hardback, *The Selected Works of Sean
Dorman*, now gradually finding its way into national and
university libraries throughout the world.

For GEORGE and JANE SULLY

Sean Dorman

Also by Sean Dorman

BRIGID AND THE MOUNTAIN
a novel

PORTRAIT OF MY YOUTH
an autobiography

RED ROSES FOR JENNY
a novel

THE MADONNA
a novel

THE STRONG MAN
a play

and published by The Raffeen Press

Under Composition
SEX AND THE REVEREND STRONG – a novel

.

Further copies of
PHYSICIANS, PRIESTS & PHYSICISTS
may be obtained through
WH Smith and other good bookshops.
Also from The Raffeen Press
Union Place, Fowey, Cornwall PL23 1BY.

PHYSICIANS, PRIESTS & PHYSICISTS

Sean Dorman

THE RAFFEEN PRESS

Cover illustration: MADELEINE DORMAN

PHYSICIANS, PRIESTS & PHYSICISTS
A RAFFEEN PRESS BOOK 1 9012430 0 1

PRINTING HISTORY
In paperback: *Physicians, Priests & Physicians* 1993
Included in three-volume hardback
The Selected Works of Sean Dorman 1993
Raffeen Press reissued in paperback 1996

Printed and bound in Great Britain by Short Run Press Ltd, Exeter.

PHYSICIANS, PRIESTS
& PHYSICISTS

The most potent reason for the production of *Physicians, Priests and Physicists* arose from the existence of my magazine *Commentary*. This monthly appeared in Dublin over the earlier part of the forties, during five and a half years. At an average of perhaps two thousand copies a month, it was a certainty that copies still lurked in collections both public and private, even possibly in newspaper files, there to haunt me. In my youthful pugnacity, had I from time to time overstated my case and fallen into folly? If I had, did it matter? Doubtless not to anyone else — but it did to me. The only way out was to republish my essays, or editorials, with inserted toning down remarks where such seemed needed. Two, of lesser consequence, I have omitted. *Good God?* was written in the sixties. The eighties gave birth to: *Danger — Doctors Working Overhead, Some Later Thoughts on Health*, and *Some Later Thoughts on God*. *Some Later, Later Thoughts on God* appeared in the nineties. An additional article, *Some Latest Thoughts on the Universe*, was added in the August of 1996.

Literary Censorship — The Supreme Impertinence

For five and a half years, during the first half of the 'forties, I edited and published in Dublin a monthly magazine *Commentary*. Apart from serving as a vehicle for my own work, *Commentary* was devoted mainly to theatre, cinema, art and music. To quote from my book *Limelight Over the Liffey*, '*Commentary* was destined to pass a lively five and a half years, with some of the best known names in the theatre and concert hall, Irish and otherwise, passing through its pages. Its liveliness derived too, I feel, at least in a small way, from the iconoclasm of a young man's mind — my own. The two idols at which I delighted to swing my hammer were the medical profession and, more especially, the Irish book censorship as then applied.

'Uneasy lest I might, in my green pugnacity, have overstated my case and stumbled into folly, I have recently consulted some early copies of *Commentary* with a view to making excuses. But I found that my younger self had comported himself with rather more sagacity than I supposed, had blundered into aggressive over-statements rather less than I feared, and that there was a certain exhilaration to be found in the whirlwind assaults of springtime vigour.'

Here then, with but a few reservations inserted, is one of the editorials, or articles, as originally published.

What more can be said against the censorship of literature in Ireland as it finds expression at the present time, so much has been said already, and so often? And yet more must be said, and continue to be said, until the little self-appointed caucuses of clerics and clods who sniff round our libraries seeking smut, and who would impose their hare-brained notions of morality on their intellectual betters, are sent packing back to the parochial breeding grounds from which they should never have been allowed to spawn.

Let us recapitulate a part of the present position, as it has been described many times in *The Bell, The Standard, Hibernia*, and other of our contemporaries. In 1929 an Act was passed known as *The Censorship of Publications Act*. It was, in its own words, 'to make provision for the prohibition of the sale and distribution of unwholesome literature, and for that purpose to provide for the establishment of a censorship of books and periodical publications.' It also prohibited the publication of certain law reports and indecent pictures.

So far, to most people, so good. Semi-privately, I think that any censorship at all does more harm than good, — I wouldn't go as far as that now — just as the prohibition of alcoholic drinks in the United States did more harm than good. Alcohol, taken in quantity, is a poison indulged in by nitwits, but it isn't half as poisonous as the loss of freedom of choice. Literature, such as that which could be bought off the kiosks of Paris, devoted wholly and full-bloodedly to eroticism is, taken in quantity, likewise the intellectual diet of nitwits, yet the social harm it does has probably been hugely exaggerated by the mixed Anglo-Saxon-Norman-Celtic population of Ireland, Wales, Scotland and England, for Paris remains, even for the sensitive and serious minded, as desirable a town as, say, Maynooth. — Note for non-Irish readers, the site of the college for the training of Catholic priests. —

Let us trace this setting up of state literary censorship further. The Censorship of Publications Act, 1929, proceeds to define the word 'indecent', a definition that is about as successful as most definitions. It goes on to set up a board of 'five fit and proper persons' appointed by the Minister for Justice. These five fit and proper chaps are, at the present moment, Professor William Macgennis, M.A.; Professor W.R. Fearon, Sc.D., F.T.C.D.; Denis J. Coffey, M.A., M.B., LL.D.; W.J. Williams, M.A.; and the Reverend Joseph P. Camac, B.A., C.C. In short, four dons and a cleric. Now I don't know any of these men but, leaving clerics aside for the moment, this I will say of dons in general:

I'm a simpleton where university degrees are concerned, indicative of what sheaves of examination papers successfully filled in what silent examination halls, examinations I could never pass myself, and can always be put upon, in print, by a fellow with a string of letters after his name as long as from here to Clark's Correspondence College. Nevertheless when I have actually met dons in the flesh they were, with a few exceptions, a set of dull conservative dogs, instinctively quoting Authorities, that is, gentlemen of some eminence who have lived in times preferably as remote as possible, in support of any position they wished to maintain, and by this ingenious means saving themselves that original thought for which they are congenitally unfitted; with a sense of humour that, in their unbuttoned moods, sported as lightly over the surface of a subject as a full grown bull elephant over a meadow of buttercups. Living their neat lives within their comfortable secure university walls, on their comfortable secure university stipends, they had about as much notion of the problems of commonplace un-university existence as a Trappist monk of swinging a hot number in front of an orchestra in a honky-tonk.

— This, coming from a writer, seems grossly ungenerous. Many authors have been discovered and rescued from obscurity by the devoted researches of scholars. Many have had their reputations kept alive by fresh editions of their work produced by university presses. Many have had their work lovingly studied, explained and placed in its historical context, and freed of misprints, by long hours of labour. Finally, in defence of myself, I *have* been known occasionally to stumble through the odd examination! —

Now any or all of these fit and proper dons may escape falling within any or all of the above definitions, but I decline to believe that contemporary literature, that sensitive and precious repository of wit and humour, of common griefs and common experience, of the purge of radicalism and iconoclasm, of that imaginative subtlety and incisiveness which probes beneath surface appearances not always wholesome to hidden essences not always unworthy, is safe in the general atmosphere which they must bring to their conference room. And, unless their clerical colleague is one cleric in a hundred, he is not likely to help.

Consider the names of but a very few of those Irish authors upon whom they have laid their interdict at one time or another (men of that tribe of writers to whom, as one T.D. — equals M.P. — had the grace recently in public to admit, the modern fame of Ireland is due, and not to her politicians): George Bernard Shaw, George Moore, Sean O'Casey, Austin Clarke, Oliver St. John Gogerty, St. John Ervine, Liam O'Flaherty, Sean O'Faolain, and Kate O'Brien. (They have also banned works by H.G. Wells, W. Somerset Maugham, Ernest Hemingway, Paul Gauguin, D.H. Lawrence, Anatole France, and Hugh Walpole.) They have super-imposed upon the existing British censorship, which in practice forms a first barrier, — since London (with New York) is the major world centre for the publishing and distribution of English language books — a second barrier of no less than 1,300 books, a great proportion of which are works of the first literary and scientific value.

It is not the priest nor the don who is the supreme philosopher and moralist, nor the politician who voted the censorship law either, but the artist — I shouldn't go as far as that now! — and to ask notable writers to have their work sat upon, and judged, and suppressed by clerics, dons and politicians is like asking Gulliver to submit in perpetuity to being bound down by the gossamer ropes of the Lilliputians. But Gulliver at length arose, and the Lilliputians scattered and fled. And, long after our Lilliputian priests and politicians are dead, and shrouded, and forgotten, the forms of those they suppressed in their little conclaves will rise giantwise upon the Irish horizon and throw mighty shadows across the face of their own land, as they now do across those of the wide world, and their names will be whispered reverentially by dons in common-rooms as Authorities in support of theses.

To proceed with the Act. It requires that, when considering a complaint against a book referred to them, the Board shall have regard, among other matters, to 'the literary, artistic, scientific or historic merit or importance and the general tenor of the book'. Note: the general tenor. Yet books are regularly submitted to the censors with particular passages marked in them. Lynn Doyle, the well known writer, resigned from the Board partly as a protest against this practice. Thus that passage could no longer be read with an open mind, lifted up and high lighted as it had been.

At a public meeting last year a lady, who holds a high position in the system of Irish libraries, exclaimed in an unguarded moment, 'It's the author's own fault in so many cases when he gets his book banned. If only he had not put in just one passage. *Even just one sentence*.' The general tenor! Many were not slow to perceive what the lady had said, and pounced upon her, and the lady stammered, and blushed, and tried to unsay what she had said, but she was too late. And when the lady had revealed the true nature of her thoughts, we saw revealed also the true nature of the whole moral attitude of this community to literature.

You will say, 'The lady is not the censorship board. Let the lady be ne'er so high up in the world of libraries, her opinions have no force, and therefore no great interest.' Which brings me neatly to the final subject of which I shall here treat, namely, the jackal of the unofficial censorship which sneaks close at the heels of the tiger of the official censorship.

The facts are now well known, and they point to the role which our library personnel might play in protecting our liberties — a role which they do not play. If there is anything at all to be said for the official censorship, it may be said that it draws its authority from a law, however half-baked, ratified by the representatives of the people of this country. But what is to be said for the impertinence of those small committees of clerics, not even appointed by the Church, who haunt our libraries and bookshops like vampires feeding upon the body of literature; who peer cautiously at the world of letters over the pale circle of their collars, with only one idea in their minds and that one addled; and who subsequently bring pressure upon the librarian in charge to withdraw from circulation books in many cases among the most celebrated in European culture. This pressure the librarian has seldom the courage, if he has the wish, to resist; so that, month after month, books are steadily and secretly withdrawn from circulation, books upon which their authors' reputations and livelihood depend, books which their readers are entitled to read. Other books are not born, because the same secret pressure is brought to bear upon publishers, with the same revolting outcome of spineless surrender. Thus the common law, which has curtailed the activities of authors and readers in certain directions, and *prima facie* guaranteed their rights and privilages in all others, and whose writ is supposed to run throughout the land, is circumvented and

11

intensified by little groups of presumptuous nobodies, who have dedicated their small existences to making the Island of Saints and Scholars an island of prigs and pedants.

Danger — Doctors Working Overhead

In my second article, — now omitted — I spoke of my few contributions to *New Health*, the magazine of Sir William Arbuthnot Lane's New Health Society. I quoted his assertion, 'The chief cause of all disease is toxaemia, which means poison.' I wrote that he considered that though ailments might appear at first sight to be very different, they all in fact arose from one single source, a poisoned blood-stream.

Why oh why were these sentences never spoken to my poor Margaret, when she went to consult not one, but two doctors? Oh, I don't mean at the end, when cancer was diagnosed. The subsequent operation revealed that it had spread. I mean years earlier, when she went with a pain in her groin. 'It's just old age, Mrs Dorman,' said the first doctor, meaning arthritis in the hip joint.

Dissatisfied with his diagnosis, she went again. He was away on holiday. His 'locum' was an Indian doctor, but unfortunately one trained in the conventional mainstream of Western medicine. Neither of them ever opened his mouth to say, 'Mrs Dorman, old age is dangerous.' (A number of people hereabouts have died from it.) 'Arthritis is a degenerative disease. It is essential that you become younger.' Had they said that, and I think that she would have heeded them, doubtless she would have been with me now. But all the Indian doctor said was, 'Mrs Dorman, it *is* arthritis.' Thus they both sat back with their arms folded, never engaging her in a vigorous programme of rejuvenation, never suggesting that that other degenerative disease, cancer, might otherwise be lurking in wait for her on the road ahead.

Why, then, did *I* not speak up and tell her the truth? She wouldn't have listened to *me*. Anyway, I didn't know it myself, or at least had largely forgotten it.

She gave little thought to her health, frequently declaring that she didn't wish to live to be old. But she had overlooked two things. She had overlooked that death, when he is a word, a fairly distant concept, is one thing, and death himself another. Many people (not all, of course) who welcome death when he is outside the house, turn to a yearning to live when he stalks into the room. She had overlooked that long life and good health are very often indeed one and the same, and that a short life is frequently fretted with illness. The term 'long life' has to be comparative. Some of us from a short lived stock will have tissues programmed accordingly.

What about me myself? Apart from the facts that I didn't smoke and had little liking for alcohol, I had lived, up to the age of twenty-three, no better than anyone else. Things improved in my last year at Oxford and on setting up in London's Chelsea as a freelance writing for health magazine. I took up exercises with and without apparatus, went on to a diet well above the average, and became a sunbathing enthusiast. Oh, I know now that this last carried some danger of skin cancer, and even possibly of an over-production of vitamin D in the bloodstream. But it carried with it also one benefit that I didn't fully appreciate at the time, and have fully tumbled to only recently. Inevitably my restless pursuit of the sun has also meant that I have passed an unusually long span of my life, for a pen-pusher that is, in the open air. But more of this anon.

After Chelsea, my diet gradually deteriorated, until it was only a little better than that of the generality. So it remained, until the recent shock of my wife's death jolted me back into the paths of dietetic righteousness — I hope not too late. Nevertheless I have had standing to me my exercises, my non-smoking and almost non-drinking, and my open-air life. And one other thing. Among the books that I ghosted or re-wrote during my four years as an external editor to a London publisher, was one whose substance was supplied by the Italian proprietor of a Shaftesbury Avenue hotel. It was a system of exercises in the bath.

This system was taken very seriously at the time by Dr Crowther of the London Red Cross Clinic, who wrote a tailpiece to the book. He made use of it not only in cases of polio and arthritis, but also in those of high and low blood-pressure. In treating alcoholics for high blood-pressure, he found that the exercises in the warm water brought down the pressure dramatically, though it returned again afterwards. But as the days passed and the exercises were repeated, it rose to an ever smaller extent, and finally rose no more. I have not the remotest idea what my own blood-pressure is (I haven't been to a doctor since I left school), but doubtless my bi-weekly exercises in the bath will stand to me too.

When my wife developed arthritis, not only would any advice that I might have ventured to proffer have been rejected, but I myself was possessed, to a considerable extent, with the notion that arthritis, though a disabling and painful disease, was not a killer. I had a great aunt whose hands were knotted with arthritis. It never went anywhere else, and she died in her nineties. My father late in life developed a stiffened knee- joint, it never gave him any pain, and when he died as the result of a car crash he was eighty-six.

But — might not a doctor, if dietetics were at the very core of his medical training, his main weapon in his war on disease and not merely a peripheral matter, instead of announcing to his patient the fact of arthritis as one might announce the existence of Mount Everest, have rather set about the task of removing the mountain, or at least as much of it as might be? The Taj Mahal is in India. The Pyramids are in Egypt. There is Arthritis in your Hip. I feel sure that had the Taj Mahal or the Pyramids eventually come to be regarded as eyesores, then the citizens of Agra and of Cairo would ere now have put in the bulldozers to demolish them. Had the doctor not so preoccupied himself with identifying symptoms and neatly labelling them, and instead have addressed himself to a body showing disease, dis-ease, non-ease, in a fashion designed to put it back into a state of ease, might he not have cured or ameliorated the arthritis, and prevented the development of the subsequent cancer?

I have spoken of my unplanned good luck in having spent so much of my time in the fresh air, or air as fresh as the motor car and the lorry will allow. There was a doctor of my acquaintance (in honesty I have to admit that he is dead now) who considered that all disease is caused by a lack of sufficient oxygen in the blood and tissues. Our tap water, doubtless necessarily protected with chlorine, and used as a vehicle to force-feed fluoride to us whether we will or no, is low in oxygen. This doctor advocated the use of rainwater or, since this

15

is not always easy to collect, the use of raw juices from ripe fruits and vegetables because of their oxygenated rainwater content.

He is not alone. Dr Holman of the Cardiff Infirmary has stressed the need for a big oxygen intake, feeling that its lack might be connected with the high incidence of cancer. The more oxygenated the body, the better it is able to offset the toxins introduced by polluted air and the wrong foods. Without sufficient oxygen in the blood, the residual wastes from the cells are not burned up. In consequence, they accumulate in the spaces between the cells and so interfere with their respiration. Malignant cells thrive in an atmosphere of carbon dioxide but, it seems, not in that of oxygen. Might not 'Margaret's' doctor have prescribed, in those early days, vitamin B 15 (pangamic acid) to increase the oxygen in her tissues? Or Vitamin E?

What about potassium? The nutritionist Cyril Scott thinks that one of the causes of cancer and other similar growths is a deficiency of potassium in the blood. In cooking vegetables, the potassium is boiled out of them and poured away. The water should be kept and taken as a soup. Anyone who does this knows well how rich it is in taste, and what an eye-opener it provides as to how much of value is otherwise being washed away down the wastepipe. Common salt destroys potassium and should be omitted. If a substitute must be taken, then let it be kelp powder or bio-salt. Worry, local irritations, aluminium cooking vesseks, cause loss of potassium. If this is not made up by eating unrefined foods rich in minerals, then growths or arthritis, or both, may occur. Dr Paul Kersch believes that cancer is caused by a shortage of potassium. He advocates a daily dose of potassium nitrate. Dr Forbes-Ross recommends a daily pinch of bicarbonate of potash in hot water as a precaution against growths.

What of liver damage from the eating of cooked flesh foods. Kasper Blond, F.R.C.S., considered that cancer is a degeneration of body tissues caused by such damage. He felt that, because cooking had destroyed the vitamins and enzymes, the meat was thereby rendered unfit for digestion and elimination. (At the London zoo, when food is packaged for the smaller meat-eating animals, vitamins and enzymes are added.) The undigested portions went rotten in the bowel, producing toxins damaging to the liver. But he also considered that the liver could be made to recover by the adoption of a

16

diet free from animal protein, and the substitution of whole natural grains and vegetables, supplemented by potassium nitrate and other tissue salts. Thus, he believed, the lives of those suffering from cancer could be prolonged indefinitely. Professor Enderlein of Berlin University was of the opinion that the eating of cooked flesh was one of the main causes of cancer.

Sir William Arbuthnot Lane considered that the toxic waste matter retained in the large bowel, due to the eliminative organs not being able to keep pace with the rate at which they were being produced, set up putrefactive gases poisonous to the bloodstream. Dr Hay's advice to those suffering from cancer was to eat all foods in their raw state, as far as was possible. Dr Kuhl of West Germany believes that cancer and all chronic diseases are caused by liver damage and impaired cell respiration due to an excess of pathological lactic acid. He considers that this last can be driven out only by eating fermented foods rich in nutritional lactic acid, such as sour milk, whey, yoghurt, cottage cheese and vegetable juices. (These last can be obtained, bottled in Switzerland, in health-food shops.) Dr Jung, also of Germany, believes that chronic degenerative diseases can be prevented by daily taking a little cold-pressed vegetable oil with its unsaturated fatty acids. He feels that animal fats can be combusted only in the presence of vegetable fats.

None of the above is one hundred per cent proven? In my boyhood I was for ever being informed that we didn't really know what electricity was. Since the universe appears to be made, so to speak, of electricity, and we have had, via Messrs Ptolemy, Copernicus, Newton and Einstein, to accustom ourselves to various and ever more accurate redrawings of the universe, it will doubtless be a long time before we know, to the last quiddity, what electricity is. But if we had waited for that, we should now be suffering an awful lot of unnecessary discomforts and loss of pleasure. And if we wait until someone has peered successfully into the microscopic world of the single cell, and discovered the means whereby it is converted into a cancer cell, before we do anything substantial, then an awful lot of us are going to be dead unnecessarily soon. The fact is that we do know what cancer is, enough to be getting on with.

We know that the best time to cure cancer is five years before we get it, better still ten (that is, unless we can manage twenty). In other words, we must alter our whole way of life to one of healthy living. We know that we mustn't smoke; that each draw on the cigarette passes into the body a smear of tar, as can clearly be seen if we make the draw through a piece of material. We know that the near presence of asbestos is best avoided. We know that nuclear waste does not make an ideal plaything for children. We know that cooked flesh foods are almost certainly dangerous. We know that fruit and vegetables, particularly when eaten raw, are a source of health. We know that we should not eat common table salt (sodium chloride), because it ends up in our eating too much salt altogether. We know that we should be wary of animal fats. 'The cow's a killer,' said two bio-chemists. Right or wrong, at least they said it with the weight of a long investigation behind them. When they spoke, they had in mind heart disease rather than cancer or arthritis, but disease is indivisable because the workings of the body are indivisable. We know that our digestions become less efficient with advancing years, and that we should supplement our digestive juices with cider-vinegar (or diluted hydrochloric acid) and ascorbic acid (vitamin C). We can sense that that doctor of my acquaintance, though now dead, was right when he urged the importance of oxygen and of the oxygenated rainwater to be found in ripe fruit and vegetables.

Might not 'Margaret's' doctor, instead of merely informing my wife that she was old and had arthritis, have said, 'Before each main meal, sip a little cider-vinegar in warm water and crunch up a vitamin C tablet. If you have a deficiency of the digestive acids, then you are not able to assimilate properly the calcium in your food and it accumulates in the joints. This calcium on the contrary must be retained in the bloodstream where, with the assistance of vitamin C, it may form healthy cartilage about the joints. Don't eat flesh foods. Don't drink tea or coffee, unless very weak, or alcohol. Seek mental calm and exercises in the fresh air. If such exercises prove painful, then perform them or others in a hot bath. The heat will soothe the pain. Also, due to the buoyancy conveyed by the water, each limb or part of the trunk to be moved weighs so much less.'

When the word Mormon is uttered, the minds of many of us fly to the concept of a man with many wives. Under the pressure of United States federal law, plural marriages were in fact renounced in 1890. There *are* some groups in Arizona who continue the practice, but of necessity they remain secret organisations. So the word 'Mormon' should in fact conjure up a people who prohibit alcohol, tobacco, tea and coffee, and who encourage participation in sport; in other words, who devote themselves to healthy living. I have read that, year in and year out, whether they are living in their own Salt Lake City or in the fumes of Chicago, their cancer rate is always below that of the general population.

Dr Kirstin Nolfi of Denmark recovered from cancer of the breast by living for some weeks on raw carrot juice, raw potato juice, raw kale juice and garlic, followed by some years on a diet of raw fruit and vegetables. Dr Maud Fere also made use of raw vegetables when curing herself of cancer of the bowel. She advised sufferers to take doses of diluted hydrochloric acid. This was to supplement the lactic acid which the cells are unable to produce in sufficient amounts to neutralise the caustic soda formed by an over indulgence in salt, and also to supply the chlorine needed for the production of gastric juice. Too much salt (sodium and chlorine, both required by the body) is easily taken because so many foods, including bread, are already salted to preserve them. This excess gives rise to caustic soda. The cells, in an effort to produce enough lactic acid to combine with the caustic soda, and so neutralise it, begin to multiply too quickly and a tumour forms. If the irritation of the caustic soda continues too long, the tumour becomes cancerous. She feels that enough sodium is obtained from celery and other raw vegetables; salads, especially beetroot; fruit, especially apples; and cooked vegetables if the cooking water is also ingested.

Dr Jung of Germany believes that chronic degenerative diseases can be prevented by several teaspoonfuls of cold-pressed vegetable oil daily, because it contains unsaturated fatty acids. The body can combust animal fats (butter, cream etc) only when vegetable fats are also present. Vegetable fats are found in such oils as wheat-germ, sunflower seed, safflower, corn; and in vegetable oil, margarine and nuts. Sunflower seeds are one of the very best sources because they are rich in vitamins B and E, necessary for proper use to be made of the essential fatty acids. These break up the globules of cholesterol in animal fats, preventing their clogging the blood vessels, and keep the blood thin.

The placenta is a very cancer-like growth. That chemical

laboratory, the body, appears to be able to shift its controls to allow the growth to develope, and then to reverse them at the time of birth to detach the growth and inhibit the appearance of another. I have read that some substances in the mother's urine at that time are very cancer-associated. Occasionally a mother with cancer will find that her tumour also has been suppressed, though sometimes it reappears later. Perhaps we shall be able to make use of the body's own chemistry to suppress our malignant growths.

Oh, and that doctor of my acquaintance, now dead, who went on so about oxygen (he called it 'lord of all') and who urged the use of the raw juices of ripe fruit and vegetables because of their rainwater (he called it 'sun water') content . . . I was acquainted with him only by reading about him. He laid great stress on diet and used few drugs. Although he lived a long life, he may in the end have died of old age. He flourished two thousand four hundred years ago. He was a Greek. His name was — Hyppocrates.

Some Later Thoughts On Health

Studies made world-wide show that the incidence of colon, breast and prostate cancer is markedly lower among those that eat plenty (the accent has to be on 'plenty'; most people eat some) of vegetables. Walter Troll, Professor of Environmental Medicine at New York University, went so far as to describe this discovery as 'startling'. Those vegetables whose edible parts are seeds or tubers (dried beans, chick-peas, cereal grains, potatoes) contain compounds that appear to intercept the activities of tumour promoters.

Feeding soya beans to a strain of rat that is highly susceptible to breast cancer greatly reduced the onset of breast tumours. (Elsewhere an experiment was made with a carcinogen which, when placed in the oesophagus, invariably induced cancer of the stomach. The laboratory animal was fed with soya beans both before and after, and no cancer developed.) The skin of mice was painted with two compounds known to initiate and promote cancer, but the cancer process was delayed and the number of tumours lower in the animals fed on soya beans. The National Research Council committee of the U.S. National Academy of Science, commissioned by the U.S. National Cancer Institute, recommended a big increase in the daily intake of seed and other vegetables, together with wholegrain vegetables and fruit.

Among vitamins, the chief inhibitors of cancer are A, C and E. Apes, which are of a similar size and biochemical constitution to humans, in captivity die of cancer, hardening of the arteries and pneumonia unless they receive a daily supplement of 1,000 mg of vitamin C. Dr Alec Forbes, MA, DM (Oxon), FRCP, for twenty-eight years Consultant Physician to the Plymouth Hospital, Medical Adviser to the World Health Organisation, and late Director of the Cancer Help Centre in Bristol, advises the same dose for people recovering from cancer. He also advises the daily intake of selenium. (This can be obtained from health food shops or chemists, a month's supply to a packet.) A deficiency of it is related to cancer and heart disease. In combination with vitamin E it helps the immune system to function, thus probably preventing cancer.

However, the vitamin most highly rated in regard to cancer prevention is A. Also beta-carotene, a provitamin that the body converts to vitamin A. Beta-carotene is found in such dark green vegetables as spinach, kale and broccoli. It is also found in deep yellow fruits and vegetables such as apricots, sweet potatoes and carrots. Dr Richard Adamson, Director of the U.S. National Cancer Institute's Division of Cancer Cause and Prevention, writes that he and his family aim to eat raw or cooked carrots every day. Vitamin A from vegetable and fruit sources is to be preferred to that obtained from animals. Dr Alec Forbes states in his book, *The Bristol Diet*, 'Carrots have been shown to prevent cancer.' He refers to an editorial, 'Vitamin A and Cancer', in *The British Medical Journal* of 1980, volume 281, page 957).

It appears that patients with cancer of the colon and rectum tend to have eaten far fewer raw vegetables than those without. They also tend to have eaten less of the cabbage family, both raw and cooked. This family includes cauliflower, broccoli, and Brussels sprouts. These contain a group of compounds called 'indoles'. Feeding them to laboratory mice has prevented cancers forming when a carcinogen had been introduced into them.

Dr Forbes records in his book that he has been seeing more and more people, at his Cancer Help Centre in Bristol, who were told that they were terminal cases by allopathic medical standards, getting well again, mostly without any medical treatment. Because the dietary methods at the Centre constitute in effect 'holistic' treatment, the treatment of the body as a whole (sometimes by a variety of methods), rather than a concentration upon a local symptom, many conditions other than the cancer are also found to be diminishing, if not entirely clearing up: rheumatism, piles, asthma, rashes, swollen ankles and thick legs, migraine.

He too, in line with the findings of the U.S. National Research Council, extols the virtues of vegetables in general. He points out that the length of our intestinal tract, some thirty-two feet, clearly indicates that we evolved as vegetable eaters. He contrasts this thirty-two feet with the tract of a meat-eating animal of similar size to ourselves, which is only some six to eight feet. Because of this short length, the meat is only partly digested and passes out of the animal before it has a chance to putrefy and break down into irritating products of fermentation.

I might add that, in order to slow down the passage of food in humans, so that the nutrients might better be extracted, the digestive canal proceeds in a series of pouches, whereas the tract of a meat-eating animal like a dog is smooth. It is not hard to imagine what is happening to meat in the heat of the body, on its progress down thirty-two feet of intestines, pouched the further to slow its progress. Our nearest cousins, the great apes, are largely building their powerful bodies out of leaves and other vegetable matter.

Some may point to the fact that our species, homo sapiens, though a very late arrival upon the surface of this planet, nevertheless has been around for some two million years, and in that time seems to have been not just a gatherer, but a *hunter*-gatherer. Doubtless the amount of meat available was small when shared out between the whole community, whereas the vegetation was all around and to hand. Further, the animals killed were fit animals, lean and free of excess fat. About the only fit animals killed these days are fish (unless we have polluted the water) and game. Meat, fish and poultry contain some thirteen times more D.D.T. and other insecticides than grains and vegetables.

Dr Forbes writes, 'If you have cancer, you should not go back to red meat. It contains a lot of growth hormones. Make sure that the chickens you do eat have not been filled up with hormones to caponise and fatten them . . . People in the West tend to make a fetish of eating meat and maintaining a high protein diet. This type of diet . . . may give you a high blood pressure and also cancer, because your fat intake is high if you eat meat. The meat we eat today is carefully bred, fed and treated with female hormones to make it very fat.'

Study after study by the U.S. National Research Council committee have connected high-fat diets with cancer of the breast and colon. The committee states, 'Of all the dietary components studied, the evidence is most suggestive for the causal relationship between fat intake and the occurrence of cancer.' Just compare our domestic bull, that perambulating ton of lard and begetter of fat, in battle good for a charge at a couple of blackberry pickers and little else, compared with a wild steer or Spanish fighting bull, strong in shoulder, neck and horn, but lean in the flanks and able to escape at speed if escape be called for. By contrast, our lumbering bull would offer an easy (and not particularly healthy) meal for any large predator.

The two major killing diseases in the West are hardening of the arteries leading to heart disease, and cancer. Our diet also promotes arthritis, appendicitis, gall stones, piles, adult diabetes and varicose veins. A doctor, working among an African tribe still living the hunter-gatherer life, found these diseases to be almost entirely absent. (There were casualties to the infectious diseases, but these were being progressively eliminated.) When he returned to work in Europe, the whole nature of his practice changed.

What of common table salt (sodium chloride)? Max Gerson MD, in his book *A Cancer Therapy — Results of Fifty Cases*, has, as one of the main aims of his diet a drastic reduction in the intake of common table salt, and an increase in that of potassium. (In biosalt, obtainable at health food shops, only one of the ten ingredients present is sodium chloride. Four of the others contain potassium.) In the average diet, about twelve times more common salt is taken than is required by the body. This tends to make us swollen, because most of the sodium is in the fluid outside the cells. Potassium is mostly inside the cells and is needed by them for their healthy functioning. A diet based on processed foods, which have to be preserved, is a diet containing far too much salt. Sodium is capable of displacing potassium from the cells, a process that tends to promote cancer. A high salt intake also leads to high blood pressure. Grains and vegetables are well endowed with potassium and have little salt.

Rudolf Breuss, an Austrian naturopath, claims, with letters to prove it, that he has treated some two thousand cancer patients who have got well on his juice fast regime. Fasts cause the body to burn itself up in order to stay alive. It is believed that the diseased parts are used up faster than the healthy. This principle has been pushed even further at the Karolynska Institute in Sweden. There route marches of some thirty-two miles a day for ten days are undertaken while on a juice fast. Despite the big loss of weight, the patients maintained a feeling of well-being.

What of the placebo effect? Geoff Watts, in an article published in *The Listener*, writes, 'The actual, as against the theoretical, consequences of taking a medicine are partly determined by what the patient believes it can do. If Mrs Jones, whose arthritis responds well to the branded drug 'Paingo', reluctantly switches to a rival preparation containing the same active ingredient, but in a tablet of different size, shape and colour, and called something other than 'Paingo', she may perceive her joints as becoming more stiff, and her pain growing less bearable. Mrs Jones's loss of faith in her treatment has, to a degree, undermined its effectiveness. The untidy parcel of psychological influences which all medicines have on their takers is known as the placebo effect.'

He speaks of those doctors who find it humbling to be reminded of the extent to which their patients are, in effect, healing themselves.

Dr Cecil Helman tells of such an occasion. A medical 'rep' was visiting him to extol the healing properties of a drug used to treat gastric ulcers. The rep displayed a chart which showed that seventy-five per cent of patients on a certain trial had had their ulcers healed by the drug, whereas only forty-one per cent had been healed by a placebo (i.e. dummy) preparation. But what Dr Helman found remarkable was that no less than forty-one per cent had had their ulcers healed by a completely inactive preparation.

It seems that in suggestion under hypnosis, and even in auto-suggestion, we have a further powerful agent in countering disease, whether it is a pill, a hypnotist, or a pilgrimage to Lourdes!

Good God?

I ended the second article in the original series with the words, 'The conclusion of Mr Hanson's article amounted to a rejection of my contention that the ideas of the atheist must be treated with respect. I am not quoting it here, or replying to it, because in a fourth article, written many years later, I took up, and took on, the theistic positions one by one. Here is that article.

God's in his heaven, all's right with the world. But is he? And is it?

If there's a Friend for little children, above the bright blue sky, couldn't he manage to be a little more friendly towards them down here also than he sometimes appears to be?

All God's chillun may have rhythm. But is it always a movement in chime with the music of the spheres? Or have they, what with all these nuclear and bacterial manifestations of God's friendliness, just got the shakes?

Is there any evidence of the existence of a god within the Christian definition, that is to say, of a god who is, among other things, benevolent and personal, computing the numbers of the hairs on our head, and looking down in tenderness upon the fall of a sparrow?

Perhaps because of the paucity of evidence to this effect, the clergyman who preached a recent sermon broadcast from the late Sir Winston Churchill's parish church, showed some signs of strain. God, thus the burden of his theme, can work only through human hands. So he made use of a Winston Churchill to bomb out of existence the dark satanic mills of the land of Hitler (and Luther). He contrasted this with the recent death in the parish of a little gypsy girl, through neglect. God could find no hands willing to help her. Seems a little hard on the gypsy girl.

It makes one think of the recent atrocities in the Congo. There was that one of pouring petrol down a man's throat, then slitting up his belly and setting it on fire. If the report be factual, then it seems that the Devil could find hands to pour down the petrol, but God could find none to send down a chaser of water. One contrasts the horrible death of this human being, with the pleasant exit of the man who, laughing over a book of schoolboy 'howlers', suddenly died from an instantaneous heart failure. What a way for a kindly and just god to run the universe! I wonder how many votes one of our political parties would get at the next election, if they governed the country like that? They'd be, very properly, out on their ear as vicious and incompetent.

If God be omnipotent, why not wave his wand and have everybody cosy? Why not have us all eat vegetables, instead of having animal to feed animal in an agony of terror and pain? Because, if God so waved his wand, he would be left with puppets, not creatures? How much pleasanter an agreeable puppet than a belly-ripping monster! Because God, although omnipotent, can't disobey his own rules? God is omnipotent, that is, he can do everything, yet he can't do everything. How the ecclesiastical philosophers love that one! It will keep them happy all summer, like children paddling.

We can't see the whole of the circle, the design, and the necessary adjustments will be made in heaven? There is not one jot or tittle of evidence for this. The whole thing smacks of wishful thinking. At what point in the slow gradation from mineral soil elaborated into the bodies of plants, from plant material elaborated into the bodies of animals, of sponges, reptiles, the lowest of the mammals, rising to the apes and dolphins, and so to man with an I.Q. of almost nothing up to a genius — at what point upon this scale is an everlasting soul supposed to have appeared? What are all these souls down the centuries being pumped into heaven for? So that God, sitting cross-legged on the carpet of the universe, like a small boy sitting in front of the fire building up and knocking down his bricks, may carry out some supposed grand design, building up stars and having them explode and break down again — so that, in short, he may play his Universe Game? God forbid that we should be saved from our plagues of the past, our cancers of the present, or our nuclear destructions of tomorrow, if it results in breaking up God's

Universe Game. Almost as bad as breaking up a game of Bridge at the Athenaeum.

We must have faith? Even the most faithful seem unable to avoid certain furtive back glances at reason. There was that pamphlet, put out by the Jesuits, that I found some good many years back left lying in the chapel of University College, Cork. Although, trumpeted the pamphlet, faith is the chief basis for belief in God, nevertheless, just in case one found oneself arguing with a non-believer — and here it gave a resume of the usual arguments. The canon, who roundly declared that he didn't think that any of the arguments had any validity, nevertheless, when very kindly writing to me on the death of my parents to offer comfort, taken unawares, I suppose, broke into reasons. Scratch one of the faithful, and you flush out a rationalist! It is not so, of course, with all of them. It is not so with Bishop Tennant, or Bishop Gore. They make no bones about combining faith with reason. 'Our religious nature,' writes Bishop Gore, 'cannot be secluded in a watertight compartment from our scientific or rational nature.'

The hundred-per-centers boost their reliance on faith by pointing out that all scientific investigation entails some element of faith. This is true. But degree matters. For example, a minor force or aggressiveness is likely to produce a counter force or aggressiveness; a policeman is taught that an angry policeman makes for an angry crowd. But an overwhelming force, as represented in a stable society by the judiciary backed by the police backed by the army, tends in the main to produce peace. So must one also make a distinction between the faith of some bird-brained fortune-teller's dupe in the presence of horoscopes or cards or crystal balls or tea-leaves; or even that of certain naive sermons that one hears preached, based on a word-by-word belief in the farrago of credulous-peasant stories that make up so much of the Old and New Testaments; between these, and the degree of faith exhibited by the scientist, always concerned as to the reliability of his instruments, always suspicious as to the reliability of his own brain, always anxious to double-check and treble-check his results.

Leaving on one side the burblings of the woolly-headed, what are the reasons for belief in God advanced by more cogent thinkers? What are they worth?

The Cosmological Argument looks at the cosmos, the visible universe, and denies that it is self-explanatory. It does not contain its complete explanation within itself, but points beyond itself.

This is assuming far too much. New knowledge is coming forward so fast, that the physicist Fred Hoyle, in a TV talk, although he had been in almost constant touch with Greenwich and Mount Palomar Observatories, had had insufficient time to work out the significance of the latest findings on the radio stars. With the advent of computers and other instruments yet to be devised, information, in such future as is left to mankind, perhaps millions of years if he does not first destroy himself, will pour in at an ever-growing speed. No man knows enough about the universe to be able to say whether or not it is self-balancing. What we know, about what bit of the universe we can perceive, *might* suggest that it is not a self-sufficient unit. But total knowledge of the total universe might well make plain that it is.

Our very sense of temporality, change, and decay, continues the Cosmological Argument, has meaning only if the eternal and the unchanging is its background. We cannot confess that all our knowledge is relative, without thereby betraying our belief in an Absolute which alone gives meaning and measure to relativity. Our knowledge of any event in nature is complete, only when the full reason for that event is found in an Ultimate which is its own *raison d'etre*, and which, because it does not depend on anything else, is not of nature but above it.

But why suggest that this is God? Because you have to call it something? Then, since a thousand ideas extraneous to this argument have clustered, down the centuries, round the word 'God', I suggest that an honest man goes out of his way to use another, say, Full Knowledge. We associate the word 'God', spelt with a capital, with sublimity and love, but the *raison d'etre* of the Ultimate might be totally uninspiring and impersonal. For a system to be eternal, it is by no means necessary that it be unchanging in its detail. A block of granite may not be eternal, but it is durable, notwithstanding being composed of a very fever of circulating particles. Certainly our knowledge of any event in nature is complete, only when the full reason is found. But it is an unwarrantable assumption to go on to say that the reason must reside in an Ultimate. It all arises from this determination to view existence as a straight line, instead of as a circle. So the straight-line boys to haring back along the railway track, their teeth shining and their eyes a-pop, to find the terminus station beyond which there is nothing. Faster and faster they go, until one asks oneself anxiously what is to become of them if in fact there is no terminus, if in fact they are travelling on a circular track, and finish by disappearing up their own backsides. In short, the universe may be viable, self-sufficient, a round-and-round affair. Its *raison d'etre* need be no more than the happenings perceives in it.

The Teleological Argument concerns itself, not with the origin and ground of all becoming, but with its purpose and end. Thinking men and women cannot believe that the many signs of design around us are sheer accident, having no ultimate significance. The universe seems orderly rather than disorderly, in that it is always realising ends. Is the whole process of organic evolution explicable, save on the hypothesis that such purposiveness implies creative Mind beyond all that is, yet working out its purpose within all that is?

The reply to the Cosmological Argument is also the reply to this. It might be added that mankind is not necessarily any example of a Universal Mind realising ends. We continue to find that we are much smaller beer than we used to suppose. Our days in the universe have been but as a candle flutter and, like a candle flutter, are likely soon to be snuffed out again. Such as we are may be fully explained by natural selection, as the mechanics of this become further revealed under our continuing research. It may seem difficult to accept that the complex structures found in nature could arise without conscious design. But the great variety around us has been built out of a very few bricks; out of a few particles differently arranged.

This rise and fall of structures, then, may be all there is to the universe. To ask: how?, may be interesting. To ask: why?, may be meaningless. Many of our present questions about the universe may be the wrong ones. After all, our flat-earth ancestors were asking some important wrong questions. What will happen if Drake tries to sail round the world; won't he fall over the edge? But there was no edge. What is underneath? Hades and supporting pillars? But there was no underneath. So why ask how the universe began? It may never have begun, not even if the Big Bang theory be accepted, for before the bang there had to be material to explode.

The Rational Argument for the existence of God says that the most significant fact about the whole evolutionary process, is the evolution of mind which is able to know that process, to think about it and to evaluate it. The very possibility of science depends on the fact that nature answers to our thought about it, and that our thought answers to nature. We cannot help believing, therefore, that the system which thus responds to mind, is itself the work of Mind, a Mind which is infinite and universal and which influences and directs the evolutionary process, because it is transcendent and creative. Such a mind cannot be contained within the universe. How and why does nature have the capacity to produce mind? Is any other answer possible, save that it does so because it never was without Mind? The beginning and the end of all things is the creative thought of God.

Again, this is almost certainly putting things back to front. Nature is as it is, and our mind has had to learn to live with it and to cope with it as best it may. (This is equally true of the other animal species, and of plant life.) Out of this adaptation of our outlook to understand the forces round us, grows the illusion that a Mind is looking back at us out of nature. But what we are seeing is not the face of God, but our own, reflected in a mirror. It is not God who made us in his own image; it is we who have made him in ours.

The Moral Argument claims that there is a consciousness of moral obligation in man which differentiates him from the animals, even the most intelligent. Man's distinctive and imperious sense of 'oughtness' has a sanctity which refuses to be bargained with, or to be explained away in terms of any alien principle. The plain man's inescapable conviction that treachery, lies and lust are wrong, is not a socially begotten value-judgement, a useful human conviction. His sense that mercy, truth and honour have eternal validity, are not the unconscious rationalisation of self-interest. An obligation wholly independent of temporal consequences clearly cannot have its origin and justification in the temporal. If this be true, then a moral idea can exist only in a Mind which is the source and sustainer of all moral excellence.

This argument creaks so badly, that I suffered considerable anxiety lest it might fall to pieces even while I was setting it down. Our so-called moral laws clearly have their origin in the temporal. They boil down to a desire to achieve a greater security, a greater comfort, a greater efficiency. Whatever may happen in the next world to a man who regularly lies or steals, in this, he is a plaguy nuisance and time-waster to his neighbours; his word cannot be taken without being checked, and goods cannot be put down without also being locked up. Our sex and marriage laws are to save us as far as possible from quarrels over mates, from the loss of our mate to another, from sexual approaches to our young before they are experienced enough to be able to conduct a mating with sagacity. Other forms of property are also protected. We are enjoined not only to refrain from coveting our neighbour's wife, but also his ox and his ass, his minicar and his record-player, and anything else that is his on hire-purchase.

We are not born with a conscience; we acquire it. We acquire it from the spoken words and the unspoken attitudes of those around us. Consider how unrestrained is the behaviour of a baby. Many of the things it does, if done by an adult, would bring the latter into court. How often we use the sentences, 'He's only a child. He doesn't know any better.' And how will he come to know 'better'? By pressure exerted on him by parents, schoolmasters, and other mentors. God working through human hands, as no doubt Sir Winston Churchill's vicar would tell us. And if there are no hands, or there are the wrong hands? Why, then the child goes to the devil. We must all concur that this is a very neat arrangement. A sensitive person may be more deeply afflicted by conscience than a stupid one. The former may, even when innocent, look guilty merely because he knows he is suspect, whereas the latter will find no difficulty in swearing that black is white. The writer and great traveller Somerset Maugham, in his *Writer's Notebook*, remarked in effect that anyone who supposed that there was such a thing as a God-given conscience, would be cured of this fantasy if he travelled more, and so discovered how enormously acceptable behaviour varied in various communities.

It is further quite untrue that the attribute of conscience, or any other mental attitude, is abruptly found in man, and never a trace of it in any other animal. Any study of animals, or listening to or reading the reports of those who study them, on the contrary do nothing else but establish how all our attributes are to be found among them in simpler form, depending on the mental development of the creature concerned. It almost passes comprehension how anyone, caring about truth sufficiently to make the effort to look at life afresh every day and question everything, could expect anything else in the light of present knowledge. Of course there is nothing to be hoped from those who, rather than surrender the beliefs of their grandfathers, would bend backwards until the tops of their woolly heads touch their heels, the while fluttering their eyelids in a frenzy of ancestor worship. They must be left to their gymnastics.

One thinks of the report of the naturalist who watched a certain species of owl in another country. The owls, borne through the dusk on ghost wings, descend silently upon the cock or hen that is their prey, sink their great talons into its entrails, and all is over. But, because of these very talons, the owls refrain from attacking one another, seemingly aware that to do so spells certain death for one. Yet, when an owl occasionally breaks the owl law and invades the hunting territory of another, the outraged tenant attacks without hesitation. The trespasser, apparently ceding that he is the guilty party, flees before the wrath of the righteous. The Ten Commandments hold sway even among owls. Thou shalt not covet thy neighbour's cock or his hen . . . Among many species there is great faithfulness to mates, even till death. The weak are often respected; females or young are not attacked. The mother seal protects her baby from the polar bear, even to her own destruction. While the baby, looking like an ice hump, stays still, the mother leads the polar bear away in a chase. She dives into the water where the bear, that magnificent swimmer, kills her. Hunters, wounding an elephant, have seen other elephants break off their flight from the guns and turn back to try to raise their fallen companion. They abandoned him only when their efforts proved fruitless. Females of various species have fostered abandoned young. Birds discharge their responsibilities in nest-building and feeding their offspring, even to exhaustion.

The Ontological Argument is an *a priori* argument, a presumptive argument, that the existence of the idea of God of necessity involves the objective existence of God. It contends that the idea of God would not enter man's mind at all, unless man's being had its source and ground in Him whose Being is wholly other than man's being and yet inclusive of it. The very idea of God is possible to us, only because God already stands behind it. Human thought is always a signpost pointing to something beyond it; deny this something, and all human thought is denied along with it. The Ontological Argument is the affirmation of faith that belief in God is an absolute presupposition of all rational enquiry.

I do not believe for one moment that belief in God arose in any such way. Man, like any other animal, wants security. When, as a child, danger threatened, he sought protection from his parents. As an adult, he sought protection from someone in his tribe who seemed stronger than he, in short, he shifted the burden of responsibility on to the tribal leader, and paid the necessary price by submitting his will to him. 'Father figure' is the modern jargon. Hero-worship, whether of god or man, comes easily to the inexperienced young, to older people of small mental strength, independence, or judgement; or to that streak in women which is specifically feminine and desires to surrender to a strong superior. It is not long before the adulation of these national leaders, whether currently or posthumously, be they a Moses or a Hitler, an Augustus Caesar or a Lenin, grows into a mystique — they become demi-gods. But there are times when even demi-gods may fail, as in incurable disease or vast natural disasters (which themselves, before their mechanics were understood, seemed to be the work of beings more powerful even than the demi-gods). Then we must raise our eyes even higher than to the thrones on which we have propped up our demi-gods. We must raise them to the highest structures we know, to the hills, to Mount Sinai, to Mount Olympus, to the Mount of Olives — to the gods themselves. In our terrible distress, we create the gods from whom alone we can now hope for salvation. When, as a child, a dog ran at us, we put our hand into our father's hand. When the doctor diagnoses terminal cancer, we put our hand into the Hand of a fantasy Father in the Heavens.

The Human Argument for the existence of God springs out of the religious experiences and acts which are integral to the history of humanity. Human history provides that basic religious experience to which the successive self-revelations of God make their appeal, and without which God's revelation in Jesus Christ would be impossible. The history of religion is not in itself revelation. Revelation is always more than religious experience. It is the divine criticism and transformation of religious experience. Man's whole religious history may not be explained as a mere outflow of his unique human nature. For man's nature is rooted in the mystery of his freedom to transcend his nature. Our history is not determined by our nature; it is not the mere product of natural necessities; it is the creative context of God's living word to man.

All this seems to me a contradiction in terms, or at least an unwarranted assumption. If a man does a certain thing, it is not a case of transcending his nature. What he has done is within his nature. He may have given a performance beyond the performance of which he is normally capable, but it is part and parcel of our nature, under extreme stimulus, to do better than our best. In this attribute, the human animal is not much better endowed than other animals. Joan of Arc, under the inspiration of her Voices (schizophrenic?), gave a creditable military performance. But our neighbour's cat didn't do badly the other evening, when crossing our yard with a dog on her tail.

The attempt to fob off the mysteries of the universe, that is, those aspects of it which have not yet yielded to our continuing investigations, on to a conception which we name God, is to achieve nothing at all. Even as we relieve ourselves of the difficulties of understanding the nature of existence by invoking God, so at the same time, in the same proportion, do we increase our difficulties in understanding the nature of God. We have merely shifted the weight from our right hand to our left. To furnish our children with computors, rather than with Bibles, would do more to hasten the coming of the Kingdom of Truth.

Some Christians are fond of quoting from the first verse of the fourteenth Psalm in the Bible, 'The fool hath said in his heart, there is no God.' Perhaps this truculent and abusive pronouncement should be rewritten to read, 'The fool hath said in his heart, there is indubitably a God.'

Some Later Thoughts On God

How much easier it was for Man, that last-minute arrival in the slow developmnt of life, having attained to a brain large and intricate enough to be capable of abstract reflection, to construct a god or a number of gods in close personal relationship with himself! How different, how smaller and cosier, was the universe that he looked out upon then, than is the universe which he must look out upon now!

He lived in a world substantially flat or lense-shaped, and not much larger than the extend of his furthest hunting expeditions. It was the centre of everything, as he was the centre of everything upon it. Whatever he didn't fully understand was a god, or emanated (thunder) from a god, or was the dwelling place (a mountain, a stream, a tree) of a god.

The sun, deified in most early cultures, was the provider of heat and light. The moon, so much gentler and therefore clearly feminine, a goddess in fact, also provided light. The planets too were gods and goddesses, or at least perfect polished orbs pushed along their paths by angels. Since Man was perceived as a creation totally different from the animals, the cruelty of the 'food chain', with creature living upon creature so often in pain and terror, was largely disregarded. Rather was it a matter of a beneficent Creator placing these lesser creatures upon the earth for the use and feeding of Man.

But, as one saw things in terms of one's own human nature, there was that other side to the gods. Clearly they could on occasion be vengeful, spiteful. 'I the Lord thy God am a jealous God, visiting the iniquity of the fathers upon the children unto the third and fourth generation . . .' Thus the King James Bible in *Exodus*. So one was reduced, in these circumstances, to suggesting to them obliquely that they *ought* to be beneficent, a bit like saying to a dog baring his teeth at you, 'There's a good dog!' Or like the Irish fairies in William Allingham's poem (the emphases are mine): 'Up the airy mountain — Down the rushy glen, — We *daren't* go a-hunting — For *fear* of little men. — Wee folk, *good* folk . . .'

This then was the believed environment in which the religions began. But what is the real environment, now known or reasonably guessed at (though a huge amount remains still be be learned)? The world is but one of many planets revolving round a medium sized star. This star is but one of millions comprising a galaxy. As far as the machine-aided eye can see, there are other galaxies stretching away in every direction to form a universe. A recent computer-based mock-up showed the universe to be balloon-shaped, doubtless at the moment expanding, but nevertheless finite. How many other universes may there not be occupying a limitless space?

Neither the earth nor the sun were recent creations. The sun appears to be some five billion years old, with an expected life of a further fifteen billions. The sun, the moon, the planets are neither gods, not goddesses, nor polished orbs; but just minerals, ice, gases or, in the case of the sun, a nuclear furnace. We are but animals ourselves. Some eighty per cent or our genetic instruction, so fantastically miniaturised as to carry much more information than the whole of the *Encyclopaedia Britannica*, is the same as that of the apes. The early foetuses of a wide range of animals, including the human animal, are close to indistinguishable. War, cruelty, the contests for mates or territory or food, and the necessary mental machinery to carry out these activities, are none of them man made. The food chain, the need for competition and conquest, are sewn into the very fabric of life. If there is a creator, then it was this same Creator who so created and so chose. Perhaps the *Litany* needs rewriting. 'O Lamb of God, that taketh away the sins of the world — first look well unto thine own.' A benevolent and personal diety? Up the airy . . . There's a good dog . . .

Mass and energy are but two expressions of the same thing. Nothing has any particular length; it depends on how fast it is moving past you. If it were able to pass you at the speed of light, for you it would have no length at all. Nothing has any particular weight; it depends on how fast it is moving. If it were able to move at the speed of light, it would have infinite mass (and infinite power would be needed to accelerate it to that velocity). Time doesn't pass at any particular speed; it depends on how fast you are travelling. The faster you travel, the slower time goes. If you could travel at the speed of light, then perhaps, for you, time would stand still. But nothing can travel at the speed of light, though it can get very near to it. (There is the theoretical possibility that particles could travel faster than light, though not at the same speed as it. But such particles have never hitherto been observed.)

How is it then, in this later intellectual climate, that so many of the old beliefs can persist? Ancestor worship seems indigenous in the human race. We revere old reputations. We go to historical plays and read historical novels. We preserve old buildings, often when they are not particularly beautiful. One of these days we'll be drooling over those lovely old Queen Elizabeth II electric pylons straddling the countryside, and those quaint old petrol pumps that are one of the glories of our towns and villages.

What of religion? What of Christianity, the most widespread faith in the West? What of its belief that Jesus of Nazareth was God, that he was miraculously born of a Virgin, that he was resurrected after his death?

In the earliest extant post-resurrection literature, the revelation of Jesus as Son of God arises only from his death and subsequent resurrection (in *Romans*, chapter one, verse four). In the *Gospel of Mark*, chapter one, verses ten and eleven, behold we find that the identification of Jesus as Son of God has been pushed back to occur at his baptism. But in neither this Gospel, nor in Paul's *Epistles*, is there any reference to Jesus's birth or infancy. This extraordinary birth, and yet never a word about it! Rather strange. The only reference is in *Galatians*, chapter four, verse four. 'But when the fullness of the time was come God sent forth his Son, made of a woman, made under the law.' There is nothing in this verse that presupposes a knowledge of Jesus's origins such as is contained in the *Gospels* of Matthew and Luke. The expression, 'made of a woman', was one commonly applied at that time to humans as a whole, and cannot be interpreted to refer to Jesus's lack of a human father.

So, taking the generally accepted chronology of the *Gospels*, we see a gradual creeping back of the identification of Jesus as

Son of God* to his birth, and beyond even that to before the beginning of time. Thus in Mark the revelation of Godhead occurs at Jesus's baptism; in Matthew chapter one verses eighteen and twenty, and in Luke chapter one verse thirty-five, it has been shunted back to his birth; while in the *Gospel of John*, chapter one verses one to fourteen, he has become the pre-existant logos (the Word of God, the second person of the Trinity), the heavenly Christ who became flesh.

Moreover, a comparison of the birth narratives in Matthew and Luke lays bare such important inconsistencies and contradictions as to leave in doubt their historical accuracy. Against some eleven clear agreements, there are disagreements almost beyond counting. Additionally, there are the internal disagreements within each Gospel itself. That extraordinary star, defying all observed gravitational forces, that guided the Magi to the Heavenly Babe, proves impossible to trace in any independent records. After all, the birth of Jesus wasn't very long ago. Less than two thousand years have elapsed, a nothing in the two million years of homo sapiens's existence as an established separate species. The observation of the motions of the heavenly bodies, even to great accuracy, had then, and long before, been an abiding interest. Aristotle was observing the phenomena of the heavens over three hundred years earlier — that same pagan Aristotle whose views about the universe were taken over by the Christian church lock, stock and barrel. And there were others, after and before. Yet not a word, not a whisper, from anybody outside that small Jewish-Christian Church. Most odd.

In short, to adopt a biblical phraseology, for those with ears to hear and eyes to see, it is obvious, from a post-Resurrection Christology that starkly contrasts with a total lack of any support in earlier literature; and from the intrinsic unlikelihood of the narratives of Matthew and Luke, that they were the compositions of a post-Easter Jewish-Christian community anxious to build up the status of their founder, hero and intellectual leader. Perhaps they

*The expression, 'He is the son of God', was generally applied, at the time, to anyone who was considered to have lived a life of outstanding goodness.

were particularly sensitive about his birth. Yet if he had been born out of wedlock, a subsequent marriage by Joseph would, in Jewish law, have legitimised the birth. When other sources are examined, the only references to Mary's virginity and Joseph's non-biological parentage all appear to be additives or alterations by the evangelists. The Virgin Birth is the product of a society whose emotional interests lay in increasing, by elaboration and invention, the authority of Jesus, of their Christ. If this sort of process were to be discovered in any other field of human activity, in politics, in industry, in science, then it would be given its true description — chicanery by stealth.

What of the Resurrection? Any attempt to reconstruct the events of Easter from extant literary sources is bedevilled by a multitude of inconsistencies. Both Paul and Luke mention an appearance to Peter, but Mark's *Gospel* seems to contain no appearances at all, except ones added later to the text. The location, number and nature of the appearances vary from source to source. Did the early Church have any reports other than the random tales of a credulous and unsophisticated populace?

Where did the idea of resurrection at all come from? The concept appears to have arisen in late Judaism as a response to a specific emotional need. This need arose out of the martyrdom of those who upheld the covenant under the Seleucid persecution in the second century B.C. The writer sought to rectify, in his own eyes, this apparent injustice by proclaiming, merely on his say-so, a future judgement for the punishment of sinners and the vindication of the righteous.

The *Book of Daniel*, in the second verse of the twelfth chapter, provides the first mention in the *Old Testament* of the idea of resurrection. But it is not a general resurrection. It is a resurrection solely for those people whose unjust treatment in this world presented a problem for the writer. 'And many of them' — not all! — 'that sleep in the dust of the earth shall awake, some to everlasting life, and some to shame and everlasting contempt.' The idea, moreover, occurs in the most fanciful of books. The stories of Daniel in the fiery furnace, in the den of lions, of his out-classing the professional sages in the interpretation of Nebuchadnezzar's dreams, would none of them have been out of place in the Tales of the

Arabian Nights. In general, it is a book dedicated rather to wishful thinking than to verity; a wishful thinking sought as a relief from a hard surrounding reality.

The *Book of Daniel* was purported by its authors to have been written in the sixth century B.C. But the sharp eyes of biblical scholars have perceived a host of internal evidence which pins it down to between 167 and 164 B.C. Worse still, the authors' fraud inspired other would-be apocalyptic writers to adopt the same presentation, whereby allegedly divine revelations concerning the future were ascribed to an ancient seer, written down by him, and sealed in a book until the time to which they referred (*Daniel* chapter eight, verse twenty-six; and chapter twelve, verses four and nine. Also compare *Revelations* chapter twenty-two, verse ten.) This splendid device allowed one to write down current history, pre-date one's book back a couple of centuries, and imply that it was a revelation of the future.

To make the effect even better, one presented the news not straight, but in a series of symbolic visions. In chapter seven, Daniel is granted a vision of four beasts from the abyss, interpreted as four empires in verse seventeen, who are brought under divine judgement; and 'one like the son of man', interpreted in verse twenty-seven as Israel, 'the people of the saints of the Most High', who is brought before God to be invested with his universal and everlasting sovereignty. The vision of a battle between the ram (Medes and Persians) and the goat (the Greek empire) in chapter eight, introduces the iniquities of Antiochus IV Epiphanes. All a bit like awaiting the result of the next general election in Britain, then stating that you had had a vision of a cow (Mrs Thatcher) fighting a bull (Neil Kinnock), in which the cow vanquished the bull (or vice versa); then back-dating your letter by a couple of centuries, signing it 'Daniel', and sending it to *The Times*, saying that you had just dug it up in your garden while planting your rhubarb.

When the authors (*prima facie* there are two, since just over half the book is in Hebrew, and the rest in Aramaic, but there has been deeper probing by scholars into the matter of authorship), when the writer concerned does attempt a genuine prophecy, he gets it hopelessly wrong. In chapter nine, verses forty to forty-nine, the account of the tyrant Antiochus's life, and the prediction of his end and thus the end also of Israel's tribulation, sharply parts company with history. With this same intention of affirming that the time of Israel's suffering is short, he drags up in the same chapter a prophecy by Jeremiah that Jeruselem's desolation would end after seventy years. As this length of time didn't fit the bill, he reinterpreted these seventy years as meaning seventy 'weeks of years', that is, four hundred and ninety years. This enabled him to bring it to bear on the period of Antiochus's persecution of the Jews. In other words Jeremiah too, as a prophet, had fallen flat on his face. Which does little to enhance one's confidence in the Old Testament Prophet industry.

The concept in the *Book of Daniel* of a limited resurrection merely for certain catagories of people, underwent the usual elaboration and invention that time always seems to bring about. From being only an idea to deal with the ethical problem created by one specific event, it became a generalised notion. By the time of the Rabbis, resurrection was announced to be scriptural, and denial of it was considered anathema.

We are only too familiar, down history, with this process, ludicrous, mischievous and disreputable, of first setting up dogma and then, to one degree or another, bullying anyone who does not subscribe to it. Science which, for all its occasional failures, has brought us so much more surely towards the knowledge of our condition, eschews dogma. Instead it sets itself, by observation and experimentation, and repeated checking of its results, to establish facts, and then to annunciate a theory which best seems to enshrine and explain these facts. This theory is entirely flexible, merely a summing up of a stage reached, always ready at need to be reshaped in the light of later discoveries. Dogma, on the other hand, is the device of the second-class mind to enchain the questing spirit of its intellectual betters, with whom it cannot keep pace, and whom it therefore fears. Dogma turns the Churches of Christendom, of Islam and of Judaism into, not fountains of truth as they proclaim, but suppressors of truth. The castigation of Copernicus by Martin Luther, and the seventy-year-later following by the Catholic Church down the same shabby path in its suppression of Copernicus' *De Revolutionibus*, were quite monstrous. And the bullying and subsequent house arrest by the Catholic Church of Galileo, whatever the political overtones, was an outrage. Truly and well, truly and well, has the Obstructive Priest of history earned his opprobrious title.

It is small wonder that both Islam and Judaism resolutely deny the Godhead of Jesus, as doubtless Jesus himself would have done. This divine attribute was foisted on to him by people writing some thirty to seventy years after his death. Small wonder that Islam has set its face so resolutely against any attempt by its own adherents to build Muhammed into a divine being. However, it has sailed pretty close to the wind. The Koran is to be regarded as the infallible Word of God, a perfect transcript of an eternal tablet preserved in heaven and revealed to the prophet Muhammed over a period of twenty years. (What an unsubstantiated rigmarole!)

In fact these three religions, with their Bible, their Koran and their Torah, all of which are claimed to have emanated from God (they are, of course, merely the work of men); and their rigid dogmas — these three religions, which have in their time screamed insults at one another, and even fought one another, are in fact almost as alike as peas in a pod. In the face of the steady advance of science, they are showing an increasing awareness of this fact. So they ever more closely bunch together like cattle, heads down, presenting their horns to the bright burning tiger whose ivory teeth gnaw ever more deeply into the fragile fabric of their fantasies.

Some have taken refuge from the onslaught by saying that science and religion serve two different purposes. This is nonsense. Both seek to make sense of their surroundings. 'Our religious nature,' writes Bishop Gore, 'cannot be secluded in a water-tight compartment from our scientific or rational nature.'

But — back to resurrection, the mind detaching itself from the brain. Even a plant has some kind of mind. It senses which is 'down' and which is 'up'. It sends its stems and its roots the right way, even in the case of a bulb accidentally planted upside down. It knows where the light is, and turns towards it. It judges the seasons by the duration of that light and by the temperature of the soil, and produces flowers and seeds in their turn. Are there artichokes in heaven? Are there radishes in purgatory, awaiting the final judgement? Do cabbages, which unseasonably go rotten, find themselves in hell?

In fossil remains, there are creatures whose only mind is in their spinal cord. When such creatures found that, to survive, they needed something more, they increased the number of nerve- tissue cells in a computer which manifested itself as a development at one end of the spine. The processes of natural selection moved in favour of those with the greater computer, more able to detect danger and plan its evasion, more able to find food, more able to protect its young. Down the centuries the development grows. At what number of cells does the Archbishop of York consider that this development constitutes a brain? At what point will the Archbishop of Canterbury rule that there is a mind, able to reflect upon itself, its motives and behaviour? Will an infallible Pope announce the method by which, when this aggregation of cells dies, the mind detaches itself to live on? And which mind? The one that we had when we were one year old? Or two? Or three or four or five? Then what about adult chimpanzees, the animal that most closely resembles Man?* By hand language we have communicated with them, and it seems that they have the mind of a five year old human child. Can a chimpanzee sit at the right hand of God? Or what about much later, when a human has suffered senile decay? Or brain damage from drugs or accident? The mind at what stage or in what condition?

The religious fellahs are fond of telling us that if you pick up a watch, you know that it has been made by someone. And if you buy a car, you know that it has been made by a machine, a robot, that itself has been made by someone. Always, in the end, you come back to an original creator. So the universe must have been made by a Creator. Only common sense.

Only common sense . . . Hm. Let us see.

All early cosmologies placed the earth at the centre of the universe. A chap had only to look out of the door of his hut to see that it was so. And that the sun, moon and planets were going round the earth. By looking out of the same door, a chap could see that the earth was substantially flat, or at least lense-shaped, and that the sky was a dome set above it. Obvious. One didn't have to discuss a thing like that with the wife. Not that the silly cow was interested in

*Another ape has been discovered, using a greater variety of tools.

anything except cooking up that fag end of bison and stinking the place out.

About seven hundred B.C. those Ionian people had the idea that the earth lay at the centre of a spherical universe. Well, OK. One had to move with the times. Some of them, though, later began to go a bit over the top, and make the rather far-fetched suggestion that the earth itself might be a sphere lying at the centre of a spherical universe. Certainly didn't look like that when a chap strolled out of his hut (that bison was becoming insupportable) and had a shufti around.

But then, some four hundred years later, those Greek blokes were at it. You had that character (what was his name?) Aristotle. Talked about the earth's shadow on the moon, during a lunar eclipse, being curved. Nattered on about ships, as they sailed away, disappearing from sight. About the stars, that could be seen, altering as one moved north or south. OK. There was undeniable sense in what the fellow was saying. The earth is curved. Perhaps this was better common-sense than one's previous commonsense.

Of course you'll always get the odd nut case. There was, some seventy years later, that freak Aristarchus, going off his chump and suggesting that all the planets, and even the earth itself, were travelling round the sun. Travelling round the sun! One had to laugh! No, better to stick with that Aristotle, really rather a sound chap, the sort of man one wouldn't mind marrying one's daughter. In fact, one stuck with him for well over a thousand years. Clever too, with his fifty-five moving spheres accounting for all the celestial motions. Commonsense, really. As was his suggestion that, whereas much was imperfect on earth, only perfection was to be found above, with the heavenly bodies polished orbs moving in perfect circles. Stood to reason! The gods were up there, weren't they, perched on the top of mountains and so forth?

These planets were pushed round by gods; well, 'angels' was the 'in' word when those Christians and, from the sixth century, those Moslems came along. Funny, though, that Aristotle said that when an arrow was fired, it continued in a straight line until the propelling force was exhausted, then plummeted directly to earth. Two straight lines! You would have thought that some infallible Pope, when they arrived on the scene, would have bent down, picked up a pebble, thrown it, and seen that it travelled in a curve. A parabola, that gaol-bird Galileo called it.

Well, there you are! Couldn't argue with the Popes. They had accepted Aristotle as received truth. Galileo in effect argued with them, and look where it got him! House arrest, and made to say that what was true, wasn't true. Still, old Kepler came a bit to his rescue, and put the boot in. Showed that the earth and the other planets travelled round the sun in elliptical orbits. Put an end to more than two thousand years of belief of perfection in the heavens; that perfect circular motion was the only one possible for celestial orbs. Put a final end, also, to the idea that the earth was at the centre of the universe. Seems that this commonsense was better than the two previous commonsenses.

Then there was Newton in the seventeenth century, with his theories of force, motion and gravitation. Perhaps the best chap of them all. Three centuries after the publication of his *Principia*, we could still use his principles to go to the moon, or put satellites on the planets. Indeed Einstein, when he came along in the twentieth century, seemed at first only to have put the icing on the Newtonian cake. But, as we have continued to delve ever more deeply into the nature of the universe, so have we ever more needed to lean upon the arm of Einstein.

His *Special Theory of Relativity* resolved the problems of the Michelson-Morley experiment regarding the supposed flow of aether, an aether that never existed except as a concept shown to be unnecessary. It resolved also the problems of the theory of electro-dynamics. It swept away all the concepts of space and time that had existed since the human race acquired the ability to have any concepts at all.

Space, or rather space-time (all events in the universe occur in a four-dimensional space-time), wasn't just an emptiness, but was capable of being warped by the weight, or mass, of the bodies in it. The earth didn't go round the sun because it was like a conker on the end of a schoolboy's string as he swung it round and round, always trying to fly off at a tangent, and always prevented from so doing by the sun's 'string', or gravitational pull. The earth went round the sun because of the warp which the latter's great mass had put into space-time.

Difficult perhaps to visualise, without an analogy for the brain to lock on to. But of course the analogy, like most analogies, mustn't be pressed too far. It is the well known rubber sheet analogy. This rubber sheet, like space itself, must be considered as frictionless. If a ball is rolled across it, the ball continues to travel on indefinitely in a straight line. If a small weight is placed on the rubber sheet so as to cause a sag in it, then the ball is diverted from its straight course. If a heavy weight is placed on the sheet, making a deep hollow, and the ball is rolled towards it at exactly the right speed, then on entering the hollow it is unable to escape, but circulates within the hollow. Since there is no friction to slow it down, it continues to do so indefinitely.

Einstein predicted that light (energy) would be bent in the presence of mass. From observations made at night, the positions of stars were known. Because of the brightness of the sun, the stars behind it could not be seen by day. But, during a solar eclipse, they could be seen — apparently out of position. The light from them had been bent in passing through the warp in space-time created by the great mass of the sun.

Then there was length contraction. The length of a moving object is affected by its motion. If a spacecraft flies at a very high speed past a stationary observer, it will appear to him to have shrunk in length by an amount which depends on the speed of the spacecraft. (The word 'stationary' has to be used with caution, since everything in the universe is in motion in relation to something else.) The closer the spacecraft approaches the velocity of light, the more pronounced this effect will become. If it were actually able to move at that velocity, it would doubtless be observed to have no length at all. The crew of the spacecraft would be unaware of this, for the contraction would have affected everything in the spacecraft in equal proportion. All that they would notice was that the *observer's* spacecraft had shrunk in length, for his speed relative to them would be the same as their speed relative to him.

Then there is time dilation. Einstein has proved to be right about this also, surely against all commonsense. Or perhaps it is just the latest commonsense that is better than all the previous common-senses. The rate at which time passes in a space craft, is slower than the rate at which it passes for a 'stationary' observer. If this observer could see the clock in the spacecraft, he would find it running slower than his watch. As the craft accelerated, the difference would become greater and greater. If the spacecraft were able to move at the speed of light, he would doubtless find that its clock had stopped.

This would affect everything on board, including atomic time and even the biological processes of the crew. If this last were not the case, then the principle of relativity would be violated. The people on board would be able to notice, after a lengthy period, that they were ageing faster than the passing of time as recorded on board would lead them to expect. But the relativity principle requires that everything on board should appear to them as normal. To the observer on earth, however, the crew would be ageing more slowly than himself. If the spacecraft were travelling at just under the speed of light, the difference in the rate of ageing would be immense, something like seven times more slowly than the ageing of the spectator on earth. So, if velocities of this order could be attained, space travel over immense distances would become possible. There would be no need for the freezing of the bodies of the crew, to obtain suspended animation, so beloved by the writers of science-fiction.

The time dilation effect has been confirmed in many experiments. In 1971 two jet aircraft, each carrying an atomic clock, flew round the world in opposite directions. On their return to their starting point, the times recorded on their clocks wre compared with the time shown on a synchronised stationary atomic clock at the United States Naval Laboratory. The result was completely in accord with Einstein's prediction.

Again, cosmic rays, that is, charged atomic particles from space, strike the atmosphere and produce short lived particles called muons. These muons, when at rest in a laboratory, decay in about two millionths of a second. In nature, they are produced at an altitude of not less than ten kilometres. Even though they travel at nearly the speed of light, they would be able to survive sufficiently long, if there were no time dilation effect, only to cover less than one kilometre. But in fact, at this enormous velocity, the time dilation factor is so large that it enables them to complete the ten kilometre journey to the surface of the earth, and there be recorded.

As far as commonsense is concerned, there's worse to come. If there are two trains travelling in the same direction, the front one at fifty miles an hour, and the back one at a hundred, at what speed would the back train overtake the front? A hundred minus fifty equals fifty miles an hour. So, if a spacecraft were travelling away from a source of light at a hundred thousand kilometres a second, and the light were following after it at three hundred thousand kilometres a second, at what speed would someone sitting at the back of the spacecraft record the light as reaching him? Three hundred thousand minus one hundred thousand equals two hundred thousand. So, two hundred thousand kilometres a second. No, it would reach him at *three* hundred thousand kilometres a second.

But that is ridiculous! Yes indeed. But it's what happens.

If a train, going at a hundred miles an hour on one track, were travelling towards a train on the other track advancing upon it at fifty miles an hour, at what relative speed would they meet and pass? A hundred plus fifty equals a hundred and fifty miles an hour. So, if a spacecraft were travelling towards a light source at a hundred thousand kilometres a second, and the light were approaching it at three hundred thousand kilometres a second, at what speed would someone sitting at the front of the spacecraft record the light as reaching him? Three hundred thousand plus one hundred thousand equals four hundred thousand. So, four hundred thousand kilometers a second. No, it would reach him at *three* hundred thousand kilometres a second.

But this is against all commonsense! Yes indeed. But it's what happens.

According to Special Relativity, the speed of light, three hundred thousand kilometres a second, measured by any observer in uniform motion, is unaffected by the relative velocity of the source and the observer. And this prediction, made by Einstein, is entirely borne out by the failure of all experiments devised to show that the motion of the source or of the observer *has* an effect on the measured velocity of light.

In view of the limited usefulness of commonsense, that is, that relatively small package of experience possessed by the computer at the top of our spine, to explain the processes of the universe, or universes; the idea that, because objects manufactured on this planet have creators, that therefore Creation must also have an independent Creator, seems to border on the simplistic, the pat, the glib. There are possibilities that it may lie within the laws of physics that something may come from nothing, and return to nothing.

A massive star, at the end of its life, collapses under its own gravitational attraction. If the final mass of the star, that is, its core or remnant, exceeds about three solar masses, then it seems that there is nothing that can halt its indefinite collapse. This will continue until the matter is compressed into a point called the 'singularity', and there crushed out of existence. However, it has to be admitted that this is not absolutely known. At the singularity, matter is infinitely compressed to infinite density by infinitely powerful gravitational forces; in other words, the space-time curvature is infinitely great at the singularity. But, it seems, we do not at present have the physics to deal with infinite situations. That must await the future.

Yet the possibility remains that something may pass into being nothing. As to something appearing out of nothing, I have heard it said that some scientists, during their experiments, have reported finding particles arriving on the scene apparently from nowhere.

In the meantime, to suppose that there are an infinite number of universes occupying an infinite space-time, which never began and which will, over all, never end, though its items may; however difficult to conceive, seems much less of a burden on the earth- bound mind than to suppose that, at some point in existence, a Creator caused all this to happen. At least, for the *questing* mind it is the lesser burden. To the tame mind, the mind that is happy to pass over its thinking to someone else, or just scarcely to do any thinking at all, the idea of an all- knowing creator is a relief. 'Mine not to understand,' it cries. 'I sink myself in Him Who is the Fount of all Knowledge.' (The tame mind always delights in capitals.)

But the questing mind has to take on a whole other cargo. This mind pervading eternity, what is it? What precisely is its nature? Why did it make up its mind that it wanted a universe, or a number of universes? Seems a lot of fuss and bother! Why not leave well alone and just have nothing going on? Was it lonely? Was it bored? If so, how did it create its toy? Just think the thought and — WHAM (or, perhaps more likely — BANG), and there was one gift-wrapped inky-dinky universe. And then perhaps, 'I think I'll have another.' BANG . . .

And now the question has to be asked, can there really be a contradiction between the behaviour of matter in large events, as described in Special Relativity; and its behaviour in atomic and subatomic events, as described in Quantum Theory? After all, both events are taking place in the same universe.

Here we might find a useful parallel by turning to Euclid's geometry. This was based on a number of axioms ('postulates' was Euclid's own word). Since Euclid's geometry was found to be rather imprecise, and also too complicated for teaching in schools, it has been the practice from the nineteenth century onwards to teach various forms of non-Euclidean geometry. As it was not known earlier that the universe was curved, it seemed sensible, in the geometry book that I learned from, to define a straight line as the shortest distance between two points. Starting from that, it could be shown that the sum of the angles of a triangle was a hundred and eighty degrees. But, since the shortest distance between two points, in a curved universe, is in fact a curve, there are many more degrees than that in a curved triangle. Does this then invalidate the results obtained by triangulation (Euclid's) when estimating the height of a tower or a tree which it is impossible or inconvenient to climb, or the breadth of a river which it is impossible or inconvenient to cross? The answer must obviously be, no. Over such short distances, the falsehood at the heart of the proposition is too small to show up. But if Euclid's geometry were to be employed to get a space traveller to, say, one of the outer planets — well, he would have to take plenty of sandwiches with him!

The situation must be much the same as between Newton's physics and Einstein's Relativity physics on the one hand, and Quantum Mechanics on the other. (Einstein received his Nobel prize in 1921 for his Photoelectric Law, and later continued to work in Quantum Theory. This latter states that energy does not exist as a continuous phenomenon, but as particles called quanta, the interaction between particles of matter and quanta of energy being explained in terms of Quantum Mechanics and Wave Mechanics.) Newton and 'Relativity-Einstein' would agree that if an experiment carried out in one place at one time gave a certain result, then the same experiment carried out under the same conditions in another place at another time would give the same result. The usual analogy used is of a man at a billiard table successfully striking a ball with his cue into the pocket at which he has aimed. If he, or another, in a different place at a different time, struck that ball with the same force and at the same angle, with all the surrounding circumstances identical, then that ball would always go into that pocket. In other words, there is predictability* present. The universe is obedient to certain laws.

In the atomic and subatomic world, however, no such predictability has been found. As a result of any given action, it is impossible to say where any particles will go. It is possible to predict only that they are more likely to go in one direction than another. To return to the billiard table, it is possible to say only, after a successful strike of the ball into the pocket, that if struck again in precisely the same way, it is more likely to go in the general direction of the pocket than in any other. It may go into the pocket, or to the right of it, or to the left. In other words, at the very heart of the universe there is a random element. (Rather surprisingly Einstein is said to have remarked, 'God does not play dice.' If this isn't dice, what is?)

However, it is a very minute billiard table and a very minute pocket, with the variations in the travel of the ball equally minute. So, like the falsehood at the heart of Euclidean geometry, the variations, while certainly present, are too small to be perceived in the context of the vast movements of the universe. Indeed, might they not average themselves out? The random element is invisible.

The discovery that the universe is finite, suggests the likelihood of other universes without number, filling an infinite space. In the matter of stars, although there is a considerable variety both in kind and state of development, yet there is a general overall sameness and repetitiveness, just as, amidst the variety of human kind, there is an overall similarity. This similarity with variety applies also to those groupings of stars, the galaxies. Repetitiveness is written across the face of the cosmos. It is likely that this applies also to that even larger grouping, a universe; that the birth, life and death (or re-birth) of

*But the findings of 'chaos' suggest that this impression of predictability has, on occasion, arisen from insufficiently accurate computations; calculations not carried on to a greater number of decimal points.

our own is being echoed in the birth, life and death (or re-beginning) of other universes without number.

As to the difficulty of intellectually absorbing the concept of a limitless space, this must be brushed aside, if only because not to do so at once sends one into the thought: if space is limited, what is outside it? Or would we, like our ancestors, be asking a wrong question? What is under the earth; supporting pillars? But there is no underneath. If Drake tries to sail round the world, won't he fall over the edge? But there is no edge. What is outside the universe; other universes? But perhaps there *is* no outside. Like energy and mass which are but two aspects of the same thing, space and time also are but two aspects of the same thing. Who knows what tricks space-time may not yet play on us?

Our forebears have had to take their share of shocks: that the earth is a globe, that it circulates round the sun, that it is millions of years old, that the planets revolve in ellipses and not in circles, that the present life forms weren't planted on the earth just as they are today, but gradually evolved over vast stretches of time; that we ourselves are an animal evolving alongside the others. The shocks do not cease, but rather seem to gather pace: that space is not an emptiness, but can be distorted by the masses of the bodies in it; that these bodies circulate round the sun or, in the case of moons, round their planets, because these distortions offer them the *only* path that they can take; that events occur in a four-dimensional space- time; that time isn't something that flows forward smoothly and evenly throughout the universe and down the aeons, but can accelerate or slow down, perhaps come to a halt, possibly reverse; that mass can become energy, and energy mass; that, even as a fish, no matter in what direction it swims, nor how fast or slow, always finds itself in the sea, so do we always find ourselves in an ocean of light or energy travelling at the ultimate speed.

What remains to be discovered? Almost certainly — simplicity. Aristotle wrestled with his cumbersome fifty-five spheres to account for the movements of the heavenly bodies. But Aristarchus made this unnecessary by his simple suggestion that the planets, including the earth, travelled round the sun. Other difficulties arose because it was assumed, on no good grounds whatsoever, that only perfect circular motion was possible in the heavens. These difficulties fell away before

Kepler's statement that the planets moved in elliptical orbits. The complications of the Michelson- Morley experiment assuming a supposed flow of aether; and also the problems of the theory of electrodynamics, both melted before Einstein's Relativity and all that developed therefrom. The genetic code for constructing a human being contains far more information than any encyclopaedia, yet it is written with only four letters. Would not all the agonising about why a loving God allows suffering and injustice in his universe, and about the creation and purpose of that universe, disappear in a simple acceptance that the universe always was here, its purpose no more than the events perceived in it; and that such a being as God does not exist?

There are few things in life more difficult to perceive than the obvious.

How To Rear A Baby

This essay, like those other essays of my early days as a writer: *Literary Censorship — the Supreme Impertinence*, and: *Mr Hanson and the Community*, was published in my magazine *Commentary* which, as I have said, appeared monthly in Dublin over five and a half years in the early forties. The same is true of the essays which follow: *The Adventure of Marriage, Marriage and 'The Standard'*, and *Jews and Gentiles*.

All these essays appeared also against the same background of public tensions. Whether these tensions specifically show in the writing or not, they affected it, because they were constantly present in the mind of the writer. The essays were produced as editorials in a magazine sufficiently widely sold in Dublin (and elsewhere) to be watched by ecclesiastical authorities who possessed the total ear of the government. This was in a Catholic Ireland much more priest-ridden (no divorce, no contraceptives), with a much more oppressive book and periodical censorship, than is the case today. What did one dare print? How far could one push? I lost my Catholic assistant editor through pressure put on him by his priest. 'I am so ashamed . . .' wrote my editor in his letter of resignation.

Perhaps the reader will think that, in my essays or editorials, I do not seem to have ventured very far in my 'pushing'. If so, he will have failed to realise the nursery level down to which the primordial Obstructive Priest had dragged us. He even invaded the art shops, as described in my later essay, *Jews and Gentiles*. In that editorial, I dared not mention that, to my certain knowledge, there were one or more priests behind the venture.

He invaded the theatre too, as described in my books, *Limelight Over the Liffey* and *My Uncle Lennox*. He went backstage during a performance in Cork of Sean O'Casey's *Juno and the Paycock*, and insisted that it should not continue in its present form. And what was the objection? The character 'Mary' announces that she could not marry her fiance, because she was bearing another man's child. Instead, the priest directed, she was to say that she could not marry her fiance because she had consumption!

At a large public meeting in Dublin, a gynaecologist complained that he was not allowed to procure from England a text book on his own subject. An Irish author failed to obtain a copy of her own book which had been published in London, presumably, she surmised, in case she might corrupt herself!

If Catholic intellectuals felt the mostly hidden but all- pervading pressure, to one like myself, brought up in the freer atmosphere of Protestantism, it was doubly stifling. Trebly so for, from the age of sixteen, I could no longer swallow the tenets of religion under any guise. Not that I disliked church- going. On the contrary, I adore it, especially to cathedrals. I adore it because I love music and singing, because I love architecture, and because I even like listening to sermons. Right to this day I listen spellbound to sermons, though sometimes I scarcely believe a word of what the fellow is saying!

One last point. Among the general public, knowledge of dietetics etc was not nearly so widespread then as now. Sunbathing, for instance, was associated with good health among the forward-looking, and hardly a thought of skin cancer around. Here then, and at last, is the article, or essay.

Before having a baby, it is as well to marry.

Marriages are made in heaven, but may split on the rocks of the Irish economic system. This system permits a wealthy man to spend on over-indulgence in food, tobacco and wine what would keep in health and comfort an entire family of the poor, the mother of which is obliged to under-nourish her children on a defective diet of the toxin, tea, and the devitalised footstuff, white bread. The rich man develops obesity, high blood pressure and early degeneration of the heart, while the children exhibit rickets, bad teeth and anaemia; whereas an adjustment of income would make it impossible for the one to kill himself by hogging, and the others by starving. This state of affairs is known as Freedom of the Individual, and is What We Are Fighting For.

Choose, then, a husband preferably not too rich, most positively not too poor. Having marked him down, if you cannot get him to chase you, chase him. Ignore any sneers of your own sex about: 'The shameless way she chases him, my dear. Of course that sort of thing only makes a man freeze up cold.'

They are jealous.

Besides, the idea that men don't like being chased is an exclusively female myth, like the male one about women wanting a good, reliable, decent man. What a woman wants is not a perfect breadwinner but a perfect lover, a Lochinvar who will ride out of the West and carry her off on a white steed, in other words, a handsome young ne'er-do-weel too shiftless to work and too stupid to learn a trade. After six months of him she will, if she's a woman of courage, cut his throat or, if a coward, her own, but in any case he's the man she wants.

But you have been warned. So choose, wisely, a man quiet because too mentally limp to be other than perpetually placid; dependable, because too lacking in courage to throw off conscience; good-natured, because too devoid of imagination to be restless; a good companion, because so without ambition that he is not for ever chained to his desk.

Nine months after you have married your Civil Servant, you will have a baby. The baby will be a boy. You will call him James (or Seamus, since the Gaelicisation of the Service).

When the nurse first brings Seamus to your bedside, you will see a little nut-brown face with, perhaps, the darker mark, which seems somehow pathetic, of the forceps on his cheek. He will be to you the most beautiful thing in the world. 'Is he not beautiful?' you will say to the nurse, and she will agree with you, reserving her true thoughts. But it would not matter what she replied, for you are not listening.

You are listening to music, to the music of cymbals, of harps, of psalteries, of dulcimers, to the music of the spheres, to the music of creation. Seamus is listening too. From that moment a secret cord binds you together, a new umbilical cord, that not all the separations of life may sever.

As the nurse removes him, you will cry out to her in an agony of anxiety to enquire if he is all right. You have seen that he is alive, but you fear that he has been born with only one leg, and that she is concealing this from you. You are satisfied only when she raises the hem of his little garment, and ten velvet-soled toes peep out.

It is indeed a relief to find that he is a complete little replica of his father; a replica, that is, physically. Mentally and morally he is, of course, quite a different being, a president, an oil king, an artist, a poet. As a matter of fact, the older he becomes the more horribly will he resemble his father, morally and mentally. But you will conceal this from yourself to the last ditch.

The matron of the nursing-home will now enter. She is a woman remarkable for her great age, and the achievement of having spent sixty years in the rearing of infants and the problems of health, without having made any discoveries whatsoever about either. Her first action will be to close the window and half pull down the blind, thus excluding both air and light. She will then sit down beside your bed and strenuously apply herself to undermining any ideas you may have had about giving breast milk (a substance developed by nature over millions of years to encourage the growth of the human young), and will urge upon you the superior merits of cow's milk (as patiently evolved by nature, for as long, for young cattle).

If you do not acquiesce, but stand out against her, she will so succeed in upsetting you that you will lose your milk anyway, so you might as well have given in in the first place.

She will forbid visitors to bring you in fruit or any other article of food which might supply material for the building of tissue, and will put you on a diet of stewed tea, remarkable for its content of the poisons tannic acid and caffeine; white bread from which a great part of the vitamin B and roughage have been removed; vegetables, from which nearly everything, including the taste, has been removed by over-cooking; and some meat, without which you would be far safer, especially while lying inactive with bowels which have yet to recover their full function.

When she has gone, and you have at length rallied from the hysterical condition to which she has reduced you, you will, being a woman of courage, insist on feeding Seamus yourself, and he will repay you with a healthy infancy and undisturbed nights.

After you have left the nursing-home, and your health has begun to recover from the shock represented by a fortnight of medical attention, you will wheel Seamus out in his pram. When you leave him outside a shop and enter to make a purchase, divers people will case admiring glances into the pram, which you will pretend not to notice; and will make adoring remarks, which you will pretend not to hear. But as a matter of fact you are living at that moment for nothing else.

That is, they will do these things if they are able to see Seamus at all. But you might be tempted to imitate all the other women and, after going to endless pains to turn him out looking as pretty as a picture, go to as many to conceal him, even on the hottest summer day, beneath a mountain of blankets, a huge hood, and gauze to ward off flies, which also have the effect of blocking out the last space through which the sun and air might have made their way. By excluding them, you will put in jeopardy the formation of his teeth and bones, his resistance to infection, and the proper oxygenation of his tissues. He will thus grow up puny, pale and unhealthy; in short, a typical and worthy citizen of his city.

Should, however, you wheel him about with the hood down, and uncovered save for a few light garments, with his arms and legs bare, you will very properly earn the opprobrium of every right- thinking woman for thus risking his death by exposure, and you will find him becoming unsuitably large, fit, and suntanned.

As Seamus grows up, you will find it increasingly difficult to get your husband off to the office in the morning, he insisting on watching Seamus take his bath, on shaking Seamus's rattle, on feeding Seamus out of a saucer and, in the case of certain Higher Civil Servants, even changing Seamus's nappy. He will spend his half holiday sitting on the floor making baby noises and playing with toys, while Seamus looks on in a detached way. You will find yourself being gradually squeezed out, and your only course in the end will be to surrender Seamus and have another child of your own.

But the secret cord which binds you to him for life will not allow you to surrender him. You will sacrifice comfort and food for him, pleasure and travel for him, you will take blame for him, you will shower love upon him and accept indifference in return, you will fight for his honour and hear no word against him, you will envisage great and romantic careers for him and marry him off to some high-born woman. But in the end a scruffy little girl will enter his life and put you out of it, he will go into the Civil Service (same department as his father) — and he will break your heart.

POSTSCRIPT

My late wife Margaret told me at the time, the forties, that some lady, with her husband in the Civil Service, asked her indignantly, 'What's wrong with being in the Civil Service?' Very belatedly I now reply, 'What indeed!'

The Adventure of Marriage

'And they lived happily ever after.'

So ended the fairy tales of our childhood, as the handsome prince led away the beautiful princess as his bride. What a wonderful faith in the tranquilising powers of matrimony!

Now, priding ourselves, as adults, on being more worldly wise, we have become cynical about the beatitude of marriage, though, as a matter of fact, we remain at bottom just as romantic and sentimental as ever. So, judging from such plays as Maugham's *The Circle*, and from the audiences which enjoyed it recently at the Gaiety Theatre — in Dublin —, we start the fairy tale somewhat later.

In *The Circle*, the curtain rises on a wife who is not at all happy in the prosperous English respectibility of her husband's home. She already has an understanding with a young man in the household. Eventually he bears her away to his shack amid the insects and heat of an African colony.

In other words, we start the fairy tale at the point where the handsome prince is already married to the beautiful princess. Only, as we are not so simple as to suppose that she is really going to be happy ever after, we show her casting sheep's eyes at another and handsomer prince, who bears her away to another and finer union. There, we feel, they will live ever after in a spiritually deeper happiness (albeit heavily qualified, on the more superficial level, by the heat and insects).

As the majority of audiences have not been married more than once, if they have been married at all, this passes muster pretty well. The people who are already starting on their second marriage, after liquidating their first, and who are already beginning to wonder whether they have not jumped from the frying pan into the fire, are still sufficiently few in number not to count.

Of course in England, where the number of divorces has jumped from five hundred to twelve thousand a year — 1945 —, it will eventually become necessary for the dramatist to start with the beautiful princess setting up house for the second time with her better and handsomer prince, with the tail-end of her eye already set reflectively on a best and handsomest prince of all.

Has, then, the beautiful princess any hope of living happily ever after? Our social radicals would be apt to say that that depends mainly on the kind of prince with whom, and the kind of palace in which, she is expected to make her life. Our Christian churches, especially the Catholic Church, would be apt to say that it depends mainly on the princess herself.

The one would rest their trust on the material plane, in labour-saving devices such as electric washing machines to remove drudgery and leave time for outside interests, and on easy divorce if man and wife thought themselves incompatible. The other, while not denying the efficacy of washing machines, or ignoring the existence of incompatibilities, would throw the emphasis upon submission to the will of God, upon self-discipline and doing one's duty, upon a quickness to judge oneself and a slowness to judge one's partner, upon waiting for one's reward in the Kingdom of Heaven; in brief, upon faith, hope and charity.

When people marry, they are usually young and inexperienced. How many of us, even with a longer experience of life, ever really know one another? It would be no easy thing for the most experienced of us to pick a partner for life. But this is the feat we are regularly called upon to perform in our twenties. Long engagements may help within limits, but those limits will not encompass the full physical circumstances of marriage. If they ignore the susceptibilities of churchmen and relatives and first live together in a trial marriage, that must tend to help, but offers no certainty. Many marriages break up after four, six, eight, ten years.

Many people marry under the impression that the only test as to whether the union is a good one is that they be 'in love'. But you may be out of love tomorrow with the person with whom you are in love today, and in love tomorrow with someone for whom you don't care a row of buttons at the moment. Many women have got into bed for convenience, and stayed there for love. Many couples have married in the very roseate aura of love, and hardly been able to pass one another the toast six months later. If they are given the remedy of divorce they will, on approaching a second marriage, be aware of this fact, and can take it into account when making their decision.

Many people marry on the basis of, primarily, physical attraction. But, as time goes on, physical attractiveness may weaken with familiarity, and the lack of community of interests increasingly count for more. The clashes will become more frequent and formidable until the moment of the final collapse.

The alternative to perpetual quarrels is an agreement to go your own ways. This may turn out to be no solution. As you ever more form your own separate groups of friends, you will ever more be apart. But marriage is being together. The less you are together, the less marriage there is. Finally the day will come when you are never together at all, and then your marriage is dead. The heart cannot live in loneliness, so you are likely to turn by degrees towards a new partner. Then your marriage is both dead and buried.

Of course you can be wrenched apart by force of circumstances, yet remain in close communion. Many loyal and loving men and women, whisked thousands of miles away from one another by the djinn of war — the Second World War — have remained in greater and more secret intimacy than couples in the same bedchamber.

Everywhere in Holy Ireland husbands and wives go their own ways in infidelities which they often scarcely trouble to disguise. This is as true of respectable people as of Bohemians, of the professional classes as the working classes, of country people as of city folk, of the ostensibly religious as the frankly irreligious.

I am sometimes tempted to divide married couples in Ireland into two classes: those who admit that they have, at one time or another, in one degree or another, been sexually unfaithful, and have; and those who deny that they have ever been unfaithful — and have. Of course I should make exceptions.

Therefore, might say the champions of divorce, since we have only the pretence of lifelong marriage here, and not a great deal of its reality, might it not be better to recognise the fact honestly and have divorce with certain safeguards? To pretend that something exists which does not exist, merely calls up the additional vice of hypocrisy, and does harm in other ways also. Prohibition in America did not prevent the drinking of alcohol. It merely gave rise to the consumption of bad smuggled liquor more injurious to health than what had been before. The pious pretence that Christian soldiers, sailors and airmen, not to mention civilians, did not go out with harlots, merely favoured the adding of disease to harlotry.

Also, it takes two to make a marriage. If one of the partners, after all the best endeavours of the other, refuses to do his or her share, what is the good partner to do? It cannot be all give on one side and all take on the other. If, driven to the last extremity by wrongs done, the good partner secures the greatest recompense which the law in this country — Eire — can offer, a legal separation, which releases him from the other but which does not permit him to marry again, why should he be condemned for the rest of his days to an unnatural half-life for something for which he was not responsible?

Yet there is much to be said on the other side, particularly if there are children involved. Is their home lightly to be broken up? Are they not loved enough to prevent that? However they are divided up between the parents, a part of their precious company will be lost. The parents' divorce will leave them poorer in spiritual credit, and in pocket. It is seldom that both partners wish to break up the marriage. It is usually only one, though the other may eventually be bullied or harrassed into it. Therefore let the law tend to throw its weight behind the home-maker, and against the home-breaker.

People do not expect their businesses to run themselves. They know that the price to be paid for success is struggle, watching and scheming, adaptation from moment to moment to meet the changing circumstances of the market. Yet some will expect that most precious vital and subtle of all businesses, marriage, to run itself haphazard. They know that the very corner-stone in any enterprise is tenacity, yet they will throw in their hand at marriage at the first emotional set-back, at the first puff of wounded pride. The Christian Church, and especially the Catholic Church, sets its face manfully against emotional half- heartedness. It will have no fair-weather sailors riding the gilded poop of the stately galleon of matrimony. It sets its face against divorce, because Jesus of Nazareth set his against divorce. He is reported to have said so, quite simply and unequivocally. If there are any ambiguities on the subject, they are man-made.

If you are a Christian, that is, one who accepts the whole teaching of Jesus as reported in the Bible, and who does not attempt to chip bits off that Rock where he finds the stone too hard, then you have no alternative but to exclude all divorce (and of course all infidelity within marriage). I myself am for marital faithfulness and for divorce in certain intolerable circumstances. But then, it follows, I am not a Christian!

Note

When the constitution of the then newly independent state of Ireland was drawn up, the Catholic Church, which had suffered much in earlier times under British rule, had a special and powerful position reserved for it. The mouthpiece for this Church was *The Standard* — or so I supposed; I had never examined a copy of it. When someone told me that I, the editor of a small magazine, had been singled out for an attack by this national newspaper, I was naturally delighted, and determined to milk the last drop of publicity out of it. Further, with the aggressiveness of youth, I was more than ready to cock a snook at self styled 'guardians of faith and morals'; to engage in an irreverent rough-and-tumble with a condescending elder and cut him down to size. Should the reader seek to discover, in what follows, anything more intellectually profound, I fear that he is likely to emerge with empty hands.

MARRIAGE AND THE STANDARD

It has been brought to my notice — December, 1945 — that I have recently suffered an attack at the hands of that notoriously anti-Catholic periodical, *The Standard*. The attack runs thus:

'A young man who edits a periodical called *Commentary* has recently made an attempt to tell Ireland what he thinks of divorce and of marital fidelity and of several other things. One can at least give the young man in question the credit for sincerity. While he sees that Christ in the New Testament did say certain unequivocal things about matrimony, he easily exempts himself by confiding to the readers of *Commentary* the terrible truth that he, the editor, is not a Christian. His readers were, I hope, duly shocked.

'I don't know any people who actually read the periodical, although I've seen it on sale when I was going to the theatre once in Dublin. It proposed to deal with theatre, art, fashion, and so on, until the editor decided to settle once for all this vexed question of marriage and divorce.

'I think I should know this young man if I ever meet him in the street, not from the definiteness of any opinions expressed, but from the size of the photograph of himself which, with the instinct of the true artist, he has filled one quarter of the area of the front page.

'He looks out from that photograph at the husbands of Ireland and tells them that they are unfaithful to their wives, and in the same breath he tells the wives that they are unfaithful to their husbands. This is how he puts it:

' "I am sometimes tempted to divide married couples in Ireland into two classes: those who admit that they have, at one time or another, in one degree or another, been sexually unfaithful, and have; and those who vehemently deny that they have ever been unfaithful — and have. Of course, I should make exceptions."

'So should I. I should even advise the young man in the photograph to move around a little, get out of the small and dizzy circle in which he seems to have lived. He might possibly find the number of exceptions so large that he would have to alter his theories slightly.

'As for those actual theories, as he puts them forward — well, I can't see them doing this country either good or harm. As somebody said somewhere: "*Commentary* is superfluous".'

The Standard fancies itself to be the most Catholic of all newspapers appearing in this country, the stalwart champion of Catholicism. But, with the above quoted farrago, it betrays the hollowness and the hypocrisy of its pretensions. For the very kernel of Catholicism, that is, Christianity, is love, courtesy, neighbourliness, but the kernel of the above article is rudeness, sarcasm, patronage. What have sneers, condescension and discourtesy to do with the gentle Jesus? — Not so gentle, it would seem, in his dealings with the money changers in the temple; and his speeches against the Pharisees! —

(It is strange how it always seems to devolve on agnostics to point out to Christians what in fact are the tenets of their faith.)

68

When first I read through the article in *The Standard*, I said to myself, 'I could drive a horse and cab through that chap's case.' Then, on re-reading it, I saw that I had mis- estimated him, and I said to myself, 'No, I could drive a coach and four through it.' Let us go driving.

The author of the article in *The Standard* signs his writing, 'Pat Murphy' — an obvious nom-de-plume. As he has not the courage to conduct his un-Christian assaults under either his own photograph or signature, I am obliged to make what deductions I can about him from internal evidence.

He does not once do me the politeness to call me by my name but, with a heavy-handed condescension, refers to me throughout as a 'young man'. (His purpose, of course, is to build up an impression of me as a bright young thing not long out of school, attempting to settle the problems of the world with a self-assured stroke of my callow pen; and making outrageous statements about my matured elders, and fancying myself very daring in so doing.) The inference to be drawn from his insistence on my youth, is that he himself is a man not young.

This not young, possibly downright old, man says that he can give me 'the credit for sincerity'. I wish I could do as much for him. Or, if he is being sincere, then he is certainly being exceedingly muddle-headed. Perhaps it is a case of senile decay. If he attributes to me sincerity, and quotes with approval my recognition that Christ in the New Testament unequivocally set his face against divorce; then how can he in the very next breath announce that 'he (that is, myself) easily exempts himself by confiding to the readers of *Commentary* the terrible truth that he, the editor, is not a Christian. His readers were, I hope, duly shocked'? Both of these sentences are loaded with the inference that I am a youthful poseur striking an attitude calculated to shock my betters into noticing me.

The phrase, 'easily exempts', further suggests that I have not taken up my attitude after much anxious and grave thought, but am merely indulging in facile mental gymnastics. In other words, after announcing in one breath that I am sincere, he announces in the next that I am not. Come, Mr Pat Murphy, you will have to try and do better than that, you know, if you hope to hold down your job. Presumably even *The Standard* expects some of its staff to think fairly clearly at least some of the time. Or are you beyond the age when that could reasonably be expected of you?

Mr Murphy makes the remark, quite uncalled for in the development of his general theme, that he doesn't know anyone who actually reads *Commentary*. This disparaging suggestion, the suggestion that some periodical of which you do not approve is not read by many people, is as outworn and as heavy-handed and as naive as the rest of his irony. As a journalist, he must know something of the very heavy cost of producing a periodical. He must know that only a steady and large circulation could support this cost; unless he presumes that I have private means. I wish I had. The plain fact is that *Commentary* is read by a very great many people indeed, from Cork to Belfast, and from Galway to Dublin, and has been read by them for four years.

He makes some impish remarks about the size of the photograph of myself which adorns my first page. If he will turn over the leaves and look at other pages, he will see that even larger photographs adorn them. (The reader will note that I have here purposely left Mr Murphy an opportunity to make a scathing distinction between my employing the verb 'adorn' to the attractive and filmic young ladies, and to my employing it about my own features. In the realms of irony, Mr Murphy can certainly be depended upon to do the obvious.)

I like large photographs. Other things being equal, I think that they are more effective than small photographs. There is a craft called 'lay-out'. An editor considers the lay-out of his pages; how to set out the pictures and the type and the advertisements on his pages to give the most pleasing effect. I like large, bold, generous effects. Certainly my photograph occupies a quarter of the area of my first page, an almost exact quarter. It was carefully designed to do so. And the title is given another quarter, and the two columns of text the last two quarters. Only, as such a pattern would, like Mr Murphy's irony, have been too simple and obvious, it was given a little subtlety and variety by the device of 'bleeding off', that is, the photograph was carried up to the neighbouring margins of the page.

As to my printing a photograph of myself at all, as an author I certainly do not seek obscurity. It has always been a source of bitter disappointment to me that I have never been able to afford to have erected, at the summit of one of the taller buildings in the immediate neighbourhood of O'Connell Bridge, a sky-sign in multicoloured neon lights, flashing out my name every ten seconds in letters some twelve feet high. Firms advertise themselves, and an author is a one-man firm. If you do not put up your own neon signs, it is certain that (except in the long run, and it may be a very long run, and you may be dead first) no one else will put them up for you. Good work that is not advertised does not bring fame. Alternatively, of course, mere publicity, without true good work attached to it, is not fame either; it is merely notoriety. But I am a good writer, and shall be better hereafter.

Therefore, I publish my photograph, and I shall continue to do so. Mr Murphy does not print his and, in consideration of the quality of his writing, the charitable explanation is that he is exhibiting a becoming modesty. Or, in view of his age, it may be discretion.

Mr Murphy suggests that I have reached my impression of the widespread unfaithfulness between married couples in Ireland, through observing the members of 'the small and dizzy circle' in which he supposes I must live. The idea that I, a non-smoker, non-gambler, and almost complete teetotaller, living most of the time a quiet life in the country, occupy a dizzy circle, would make my friends hoot with laughter. I run considerably more danger from the adjective 'staid' than from the adjective 'dizzy'.

It was because I knew that the attempt would be made to explain away most marital failure on a dizzy circle, that I made the statement, and it is true, that this failure was met with just as often among respectable people as Bohemians, among the professional classes as the working classes, among country people as city folk, among the ostensibly religious as the frankly irreligious.

That this is true, Mr Pat Murphy would know well, if he knew the people of this country at all. In fact, he obviously knows so little of them, that this, together with the studied over- Irishness of his nom-de-plume, Pat Murphy, inevitably leads to the conclusion that he is neither an Irishman nor a Catholic — probably a Scottish Methodist.

This elderly Scottish Methodist tries to curry favour with the Irish people, by assuming the mantle of protector of the good fame of marriage in Ireland. He glibly brushes aside my observations with

71

a few unoriginal flourishes, the while smiling smugly over his shoulder at the Irish people, whose admiration and support he feels sure he possesses. That might work in Scotland, though I doubt it, but it certainly doesn't work here. The Irish people have too much underlying seriousness and sensitivity of intellect, to be imposed upon by a foreign mountebank seeking an easy popularity.

A few supporters Mr Murphy will have, of course. There will be some of the unmarried, whose notions of marriage derive out of daydreams rather than experience. There will be those natural hypocrites, born so completely devoid of the power of seeing themselves as they are, that they have the capacity of combining being no better than their neighbours, with the most outraged indignation when they are told that they are no better. There will be those people who have truly dwelt in the shade of their marriage vows, and all honour to them, and are too blindly charitable to suppose that their neighbours have not done the same.

But there will be the others whose homes have avowedly broken up. There will be those whose homes are a respectable facade desperately maintained in the faces of their neighbours, while in the cellar smoulders a slow fuse that tomorrow may ignite the explosive that will blow the shell into fragments. There will be those whose marriages have barely held together only after many a battering and many tears. There will be those who, though their own marriages are happy and secure, know well that those of some of their friends are not. I do not think that Mr Murphy's smug assumptions will hold much appeal for them. Smug is for the smug.

It is because I care so much for marriage, and am so saddened and unsettled and depressed at the signs of disharmony on all sides between men and women, as I am at all disharmony, that I was moved to write as energetically as I did. *I* shall plead no brief for home-breakers. *I* am no admirer of cynicism and disruption and animalism and disloyalty.

This is Christmas, and it is fitting that Mr Murphy and I should part as amicably as possible. And so, as he croons by the fire this Christmas-tide over his bowl of haggis, I wish him — A Merry Christmas.

Jews and Gentiles

Mr Victor Waddington was exhibiting a picture of a nude woman in the window of his gallery in South Anne Street — in Dublin, November 1944, at the height of the holocaust of the Jews by Hitler — when five people, calling themselves a committee, entered and demanded that he remove it.

The picture was a reproduction of one of the most famous paintings in the world, *Olympia*, by the great French Impressionist, Edouard Manet. It was but one of a great number of reproductions of famous Continental paintings which were taking their turn at being exhibited in the window, with four days allotted to each. The purpose of this was to link up with the Loan Exhibition of Modern Continental Paintings at the National School of Art, by far the most important exhibition this country has seen for many a long year.

On Mr Waddington's refusal to remove the picture, one of the five said, 'You are in this country as a guest (Mr Waddington is a Jew), and it is up to you to respect the feelings of the people whose hospitality you enjoy.'

Mr Waddington, whose talent and industry have made his galleries by far the best commercial galleries in the entire country, the chief meeting ground of artists and buyers; who has been accorded the most generous treatment, during the war, in the shape of travel permits to London to bring pictures over to this country, so highly do our Authorities rate his services to Irish culture — Mr Waddington replied that there was nothing which he had received from the community that he had not fully repaid. He then made this offer to his five intruders. If they would go to the National School of Art, and have removed from the exhibition there Manet's water-colour sketch for *Olympia* (his try-out, so to speak, for his

Louvre masterpiece and virtually the same picture), then he, Mr Waddington, would remove the reproduction.

One of the intruders replied, 'We are not concerned with the School of Art.' They departed.

Some days later two stalwart young men presented themselves, demanded the removal of the picture and, on being refused by Mr Waddington, said that they would break his windows if he did not do so. He replied that he had the name and address of one of the five who had previously visited him, and that he would trace the rest and charge them jointly for the damage. Nonplussed by this, the bullies left.

The four days allotted to the showing of the Manet expired, but Mr Waddington, feeling that to remove it would be taken for surrender, extended its period in the window.

Effrontery had failed. Threats had failed. Now intrigue was tired.

A little time later, while Mr Waddington was away, some people approached one of his employees and said, 'You know that picture in the window? Now you're a decent Catholic girl. Remove it, and if you suffer dismissal as a consequence, we'll take the matter to law and make it all right for you.' She replied that she was perfectly satisfied with Mr Waddington as an employer and would not remove the picture without his instructions.

Good girl! Erin go bragh!

Consider a few of the implications of this episode.

People are entitled to come to the conclusion that such and such is a good cause. They are entitled to band themselves together to watch over the interests of that cause. Two examples which spring to mind are the Society for the Prevention of Cruelty to Children, and the Society for the Prevention of Cruelty to Animals.

If their cause appears to be in jeopardy, they may take measures to come to its support through the law of the land. If the law, as constituted, offers them no redress, then they still have the various methods, guaranteed to them by democracy, of agitating to have the law amended, such as the calling of public meetings, the issuing of printed manifestos, an approach to their representative in the Dail — Irish parliament —, the use of their vote. (I suggest, however, that if anyone should desire to force the passage of a Bill for the Prevention of Artists Painting Women Without their Clothes On, he would be several centuries too late. The matter has gone too far for him to reverse. If such a man had his way, Titian's masterpiece, *Sacred and Profane Love*, would never have been painted, and some modern Rubens would be cooling his heels in the Bridewell.)

But what is to be said of this precious little group of bullies attempting to take the law into their own hands! How dare they have the impertinence to arrogate to themselves the right to represent the people of Ireland! God forbid that anyone should think that the attitude of our country to art is represented by this pack of bigots and hobbledehoys. They were ready enough to display their fine courage in the cause of virtue, by trying to browbeat a single tradesman, a tradesman, moreover, belonging to a small and often persecuted and hated community; but when it came to the big exhibition at the National School of Art, with its powerful backing, they 'were not concerned'.

The attitude is often taken: it is one matter to display something semi-privately to a select and confined gathering; it is another to do so publicly to every Tom, Dick and Harry. Greater licence might be allowed, for instance, in a theatre, where you may or may not take your daughter, or your mother, or your Aunt Sophia, as you please; than on the radio, which forces its way into every parlour. It is one thing for people, habituated to the conventions of art, to pay to enter an exhibition and there to see a nude; it is another to have this nude exhibited in a window in a busy thoroughfare where innocent daughter, untouched mother, or virtuous Aunt Sophia, might be staggered by a prospect not greatly dissimilar from one that they must be obliged to face every time they take a bath.

But even if we allow for a moment the attitude that a certain picture may be shown with propriety in one place and not in another, consider the circumstances further. The entrance cost to the exhibition was small and hundreds visited it and saw Manet's nude. I'm sure that several Toms, Dicks and Harrys, and probably even Aunt Sophia, managed to slip in. As for Mr Waddington's reproduction, it was not being shown in a small country town, but in the capital city and cultural centre of the country, Dublin. It was not being shown in a suburb of Dublin, but in the most sophisticated and fashionable part of the town, in the chief art centre amidst a cluster of commercial and non-commercial picture galleries and art shops.

The nude has been accepted in art from time immemorial. If a single nude cannot be exhibited in a fashionable art shop, in the fashionable part of the capital city, where the educated and the sophisticated dwell and shop and work, then what a nation of clods we have become.

And what of the suggestion that the Jewish community dwell in this country as 'guests'?

The Jewish community dwell in this country as Irish citizens, as Irish men and women. They are the guests of nobody; they are beholden to nobody. They are living upon their own soil. Their titles are as unchallengeable as the titles of all the O'Donovans, the O'Flahertys, the O'Gradys and the rest of the O apostrophes in the telephone directory.

It is most disagreeable to find anti-Semitism rearing its extremely repulsive head in Dublin. I come across small evidences of it from time to time. The Government remains correct and democratic in its attitude to minorities. But tolerance at the centre cannot obviate local intolerance. If this last is not fought and checked, then the central government, since it is elected by localities, will become contaminated too.

Some of the most resolute in refusing to meet Jews socially, have been among certain of the so-called Anglo-Irish of my acquaintance. If a Christian does something wrong, they allow him to have done it as a private individual. But if a Jew commits a misdemeanour, it is held to be typical of the race, and they smear the pitch of their condemnation over the whole of Jewry. 'These Jews!' they say. Why not, for a change, 'These Christians!'

Foolish people! Let them beware of denying Jews their full place in the Irish sun, lest the day arrive when the Catholic majority will say, 'These Anglo-Irish!' Perhaps they will murmur that you are the remnants of a British garrison, are guests here on sufferance, expected to walk humbly and be duly grateful for your keep. And then, when the Jews and the Anglo-Irish have both been set aside, it will be the turn of the Catholics to cast suspicious glances at one another, each questioning if the other is a true Gael and has a proper title to the land.

Liberty, like peace, is indivisible. There is no end to the career of the mad dog of religious and racial intolerance and, once he has been loosed over society, who shall chain him again?

Some Later, Later Thoughts On God

The preceding essays, including theat entitled *Some Later Thoughts on God*, were, for the most part, set up in print a year or two ago. This came about because they were originally intended to be part of a book called *The Strong Man*, the title of a three act play which started it off. That original book was an uncomfortable mix. The theatrical piece was accompanied by a series of essays which had nothing whatsoever to do with the theatre. When in addition it began to grow over-long, I joyfully suspended the printing and brought out the play, bolstered by some theatre critiques, as a publication on its own.

As said, *Some Later Thoughts on God* was printed a year or two ago, but was written quite some time before that. I have since, rather belatedly, come across that fascinating book, *A Brief History of Time*. Some of the remarks made in it by the great mathematician (of Cambridge, unfortunately; why hadn't he the sense to work in Oxford!), Stephen Hawking, have caused me to feel uneasy. Hence, *Some Later, Later Thoughts on God*.

In the original essay, I guessed at the future lifetime of the sun as being three times as long as its past. I see that Stephen Hawking's approximation gives it a future of the *same* length. Indeed, I have seen this estimate suggested elsewhere also.

I wrote of 'that pagan Aristotle whose views about the universe were taken over by the Christian Church lock, stock and barrel'; and did not, it seems, give enough credit to Ptolomy for the prolongation of those views. He it was who, some four hundred years later, developed Aristotle's understandings into a simpler cosmological pattern, the earth being surrounded by eight spheres only, bearing, respectively, the moon, the sun, Mercury, Venus, Mars, Jupiter, Saturn and the 'fixed' stars. This left any amount of room outside the stars for the Scriptural heaven and hell, so the Popes could rest happy in this cosmological meadow. Unfortunately there proved to be a snake in their grass — a Polish priest, one Nicholas Copernicus.

Further on I said, 'Some have taken refuge from the onslaught' (of science) 'by saying that science and religion serve two different purposes. This is nonsense. Both seek to make sense of their surroundings. "Our religious nature," writes Bishop Gore, "cannot be secluded in a water-tight compartment from our scientific or rational nature." '

I feel that I should have elaborated on this by a reference to six German Protestant pastors who fully took on board the immense achievements of science, but who each had their different ways of harmonisings their beliefs with these achievements.

Friedrich Schleiermacher (1768 to 1834), Professor of Theology and Dean of the Faculty of Theology in the University of Berlin, with his *On Religion, Speeches to its Cultured Despisers* and his *Systematic Theology*, created a sensation. It obviously would be impossible to do justice here, except at inordinate length, to his thought. One or two clues will have to suffice. In the first work, the second speech, he writes that religion 'resigns at once all claim on anything that belongs to science or morality.' He says to the 'cultured despisers', 'You would never grant that our faith is as surely founded or stands on the same level of certainty as your scientific knowledge.' 'Faith' and 'scientific knowledge' are to be firmly put apart because of difficulties over 'level of certainty.' The authenticating knowledge of God is not derived but direct. If God exists, then the only person that can establish that God exists is God Himself.

Rudolf Bultmann (1884 to 1976) was Professor of New Testament Studies at Marburk from 1921 to 1951, and is famous for his demythologising of the Bible. A notable example of mythical thinking was the Virgin Birth of Jesus. Such a mythical approach to reality, he believed, was quite out of the question for modern man. As a *cause*, God is out of the universe; the universe gives no sign of his existence. The liberating awareness of authentic existence comes by faith alone. God is found solely in faith. Bultmann's talk of faith as a decision taken, makes it seem that it is only a man's or woman's decision to seek out God which makes it possible for God to reach them.

Karl Barth (1886 to 1968), on the contrary, would consider that this was blasphemy; that the survival of theism depended on the fact that God continued to choose out certain people. Barth was Professor of Theology in Germany until expelled by the Nazis; and, at a later time, at Basle University in Switzerland. His *Commentary on Romans* marked on extreme severance from nineteenth century theology.

Emil Brunner (1889 to 1966) agreed in general with Bultmann and Barth about the necessity, in the face of the progress of science, of setting up the Christian position anew. Emil Brunner taught theology at Zürich from 1922 to 1938, and at Princeton in the United States from 1938. He was the author of *The Mediator, The Divine Imperative*, and many other works. He too considered that the Christian faith could be found and kept by revelation alone. Faith was wholly and necessarily self-authenticating.

Paul Tillich (1886 to 1965) wrote perhaps more widely than any. His interests ranged from Theology to Philosophy to History to the study of Cultures. This tended to keep him more closely in touch with the movements of thought in the Western world. Paul Tillich was a Professor of Theology, then later of Philosophy, in Germany. When the rise of the Nazis drove him out, he became Professor of Philosophical Theology at Union Theological Seminary in New York. Despite his existentialism, he preserved a certain measure of faith in the older assumptions of a connection between the evidence of the existence of God and the perceptions to be found in a study of the universe. He aimed to take up an existential position while accepting Schleiermacher's complete severance between Science and Religion (and thus some denial of Natural Theology); together with some part of Barth's and Brunner's 'necessity' of the givenness of revelation. But he insisted that the exposition of the understanding of God had to tie in with the questions that people were generally asking about the situation in which they found themselves.

Dietrich Bonhoeffer (1906 to 1945) at first agreed with Barth and Schleiermacher that the world provided no data about God; largely agreed with Barth that the 'Word' of God *was* the data about God. But, as time moved on, his deep involvement in the Second World

War and his resistance to the Nazis had its effect in drawing him, and with him his theology, more and more into the world. Dietrich Bonhoeffer did his teaching in Berlin until stopped in 1936 by the Nazis, angered by his involvement with the Confessing Church in Germany in their struggle against Hitler and what he stood for. Bonhoeffer had many friends in Britain and America and could have saved his own skin. Nevertheless, when war seemed certain, he left the USA to join the German Resistance. He was arrested in 1943 and hanged in 1945. Barth and Bultmann seemed to him to be neglecting a world in which he himself was so stressfully involved; to be sheltering themselves in a theological cocoon. But that world, to which Jesus had been so deeply committed, must be taken seriously. Man had 'come of age.' He no longer felt so totally obliged to await divine intervention to put things right; he could put much right with his own technology. Nor was he prepared to stomach the sermons of divines, scared of losing their power (Tillich made Bonhoeffer particularly cross), presenting this life as a vale of woe, a fiery furnace through which man must pass to clear away (assisted by the Church) the dross, leaving the gold to pass into paradise. Man had plenty of happiness here below, brought about by his own inventions. He certainly had no need for gods, or God, in order to explain the mechanisms of the universe.

In my essay, *Some Later Thoughts on God*, I made a rather cryptic reference to the Michelson-Morley experiment which both failed to explain what that experiment was and, more importantly, to attribute a proper degree of honour to them, and to Maxwell. In 1887 Albert Michelson and Edward Morley compared the speed of light in the direction of the earth's motion, with its speed at right angles to it. Astonished, they found that they were exactly the same. Maxwell's equations predicted that the speed of light would be the same, no matter what the speed of the source. When a stone, at a certain moment in time and position in space, is dropped into a pond, the resulting spreading ripples are, immediately after that time and position, totally independent of the later movements or speed of the stone. The stone, having initiated an event, no longer exerts the smallest influence on the events that follow. Similarly, when a light source emits a pulse of light at a certain moment in time and position in space, that pulse of light will continue to spread out into a cone at its own speed, totally independent of the later movements or speed of the light source.

In my previous essay on this general subject, I mentioned Einstein's Special Theory of Relativity, and neglected to relate it to his General Theory of Relativity. In 1905, Einstein (and Poincaré) were very successful with the Special Theory of Relativity in explaining the results of the Michelson-Morley experiment, that the speed of light appears the same to all observers. But this theory was inconsistent with Newton's theory of gravity, which stated that all objects attracted each other with a force proportional to their mass, and in inverse proportion to the square of the distance between them. This meant that if there were two objects, and one was removed, the force exerted on the other would change immediately; that the effects of gravity could travel with infinite velocity. But, it seems at present, nothing can travel at the speed of light (except light and other waves without a mass content), let alone faster. After wrestling with the problem for a number of years, Einstein came up with his General Theory of Relativity. This stated that gravity was not a force as such, but was the consequence of space-time being warped by the distribution of the masses in it. Bodies in the cosmos don't move in their orbits because of gravitational forces, but, rather, are merely following the shortest possible path (a geodesic) in four-dimensional space-time. (Geo, from Greek ge, earth; so geodesic, used more literally, is the shortest distance between two points on the surface of the earth, taking its curvature into account. Long-distance aircraft commonly follow a geodesic.)

It has also become clear to me that my attempt, in the previous essay on the subject, to brush aside quantum mechanics as an influence too small to effect the great movements of the universe, thus leaving the field clear for relativity, was very wide of the mark. On the contrary, the random element[1] (Heisenberg's Uncertainty Principle) is of such crucial significance, that the race is on (together with an ever more ferocious bombardment of the present smallest known particles to break them into the smallest of all, the building

[1] *Consider also the ideas of 'Chaos' and 'Anti-Chaos' — the tendency of disorder to return to order.*

bricks of the universe[1]) to marry relativity and quantum mechanics into a quantum theory of gravity; a single explanation of the universe. Given this, we should be able, within the limits of the uncertainty principle, to trace the history of the universe back to its beginning, if it had, in any complete sense, a beginning, which seems in doubt; and forward to its end, which also seems in doubt.

According to relativity, since the universe is expanding in every direction, all the material in it must in the past have been at the same place[2], the 'singularity', where the Big Bang occurred. As the force of the explosion grows ever less, and gravitational forces ever more assert themselves, the rate of expansion will slow, cease, and give way to a contraction, culminating in a further singularity, the Big Crunch. (There is the possibility that, whether within a black hole or generally, the material in a singularity may be gradually returned to the universe. This could alter the picture to one, perhaps, in which the expansion of the universe arose from a curved position, a boundary, rather than an explosive point; and will eventually contract to a curve rather than to a point, at no time requiring either a Creation or a Creator.)

If the Big Bang *did* occur, then the reverberations from it should still be around in the immensities of space-time. (Both space and time can have existence only within a universe; neither can exist 'outside'.) Now, it appears, the reverberations of that primal explosion have definitely been heard. When the material in the universe returns to the Big Crunch, might not that colossal compression initiate yet another Big Bang — with a repeat performance? And how many other universes and space-times may there not be?

In that previous essay, I spoke of the genetic code for constructing a human being containing far more information than any encyclopaedia, yet being written with only four letters. It appears, from the investigations of Professor Carlos Bustamante and Mr David Dunlap of the University of New mexico, with their scanning tunnelling microscope capable of taking a direct image of the 'letters' of the

[1]*But the Superstring Theory suggests that the building bricks of the universe are not elementary particles, but minute strings vibrating at specific frequencies and rotating in ten-dimensional space.*
[2]*or at least close together.*

genetic code, that the code is written in combinations of three of these letters, used by cells to make all the molecules necessary for life.

In 1951 the Catholic Church declared its belief in the Big Bang, since it considered it to be in accordance with the Bible. Stephen Hawking, the great Cambridge theoretical physicist (I still think that he ought to have remained at Oxford where he first was!), relates that in 1981 he attended a conference on cosmology, organised in the Vaticam by the Jesuits, doubtless mindful of the cropper the Church had come over Copernicus and Galileo. At the end of the conference, he and the others were accorded an audience with the Pope. The latter informed them that it was permitted for them to enquire into the mechanisms of the universe after the Big Bang, but that they must not enquire into the Big Bang itself, because that was the moment of Creation and therefore the work of God. Since Stephen Hawking's paper to the conference had been doing that very thing, he felt it fortunate that the Pope was not aware of the subject of his talk; indeed, in addition, that it had been couched in such highly mathematical terms, that he would not have understood it. Hawking had no desire to suffer the fate of Galileo!

He is still with us, isn't he? I mean, the Obstructive Priest and his Holy Book, whether it be the Pope and his Bible, the Ayatollah Khomeini and his Koran, some Rabbi and his Torah, blundering his way through history. (There are, of course, many reasonable priests.) With his menacing thousands of followers who, despite their protestations to be adherents to a God of Love, when roused prove indistinguishable from a mob serving the ends of the Devil. 'Thou art free to enquire thus far, but then thy intellect must be shut down.' 'Why?' 'Because I say so — *ex cathedra*.' 'But why say you so?' 'Because it is written in the Holy Book from long time ago.' 'But at that time the writers didn't know as much about the nature of the universe as any quite young well educated present day child!' 'Silence! The Book was written by God.'

'Baloney!'

END

84

Some Latest Thoughts On The Universe

Isaac Newton's mechanics, although they did not take account of friction or air resistance, both of which he doubtless considered to be negligible, held sway for hundreds of years as a true description of the laws of the universe. Then all was put in doubt by the discovery of electricity and magnetism. These showed that Newton's 'laws' could be applied only to the motion of solid bodies. Worse was to come. It was found that these 'solid' bodies were not so solid after all. They were, in fact, composed of atoms. Newton's mechanics now applied only to bodies made up of large numbers of atoms. But not always then. They applied only when these bodies were moving at speeds slow when compared with the speed of light.

When a body is composed of only a few atoms, then quantum theory takes over. When a body is moving at, or near to, the speed of light, then relativity theory rules the roost. Quantum theory and relativity theory. These two have stood up to thousands of tests and not been found wanting, quantum the theory of the very small, relativity the theory of the very large. But how to *unite* them, has proved the very devil.

All these models, Newton, quantum, relativity, give approximations which are valid for a certain range of phenomena, but outside their territory they no longer explain nature satisfactorily. So the effort must be made to extend them further, at least to give closer approximations if not complete answers.

Not content with the elementary particles whose collision tracks can be photographed in the bubble chambers of our giant particle accelerators (the best hope of finding out what is inside a particle is to break it up), there are those who are inviting us to join, at least at second hand, in the search for quantum strings that go to make up these particles; that are as much smaller than the particle, as the particle is smaller than ourselves. Millions, upon millions, upon millions, upon millions, upon millions, of these strings, laid end to end, would measure only one centimetre. Which is quite short! These superstrings are visualised as extended objects that spin, rotate and vibrate. They have the virtue of offering a view of nature that is both elegant and simple — and simplicity is often the truth.

They do, however, bring with them a difficulty. This difficulty is that the concept works best in a universe of ten dimensions. (The number of dimensions has been much argued over, ten and eleven being particular favourites, but things seem to have settled down at ten.) But we all know that our world consists of four dimensions only: side to side, up and down, to and fro, and time. So, if there were ten dimensions at the beginning of the universe, at the Big Bang, it will have to be shown how these six extra dimensions have managed to remain hidden and compactified. An enfolded shape called an orbifold has been proposed. But why should this compactification remain stable in a world of quantum fluctuations, and a universe in which other spatial dimensions are expanding?

While the theory of superstrings was evolving, Roger Penrose, Professor of Mathematics at what I like to think of as my own university of Oxford, was developing his different approach, using his objects known as 'twistors'. Like superstrings, his twistors are extended objects that have the ability to describe both the internal structure of the elementary particles and the quantum nature of space-time. Rather than needing a higher dimensional space for their origins, they are defined in a three- dimensional space, but one whose dimensions must be described with complex numbers. Complex numbers, in essence, are extensions of ordinary numbers got by including the square root of minus one ($\sqrt{-1}$), usually written, for brevity, as 'i'. These complex numbers have greatly extended the powers of mathematics.

But there are those who would argue against the superstring search for entitles even smaller than the known particles. They would say that there is no one smallest indivisible object out of which the universe is built. When two particles are banged together in a particle accelerator and break into pieces, these pieces are no smaller than the original pieces. Indeed, they emerge as particles of the same kind, created out of the energy of motion, that is, kinetic energy.

The universe appears as a web of inseparable energy patterns. Both force and matter have their common origin in the dynamic patterns that we call particles. The universe is found to be an inseparable whole. The further we push our way into the sub-microscopic world, the more it is forced upon us that we are in, and ourselves composed of, a system of inseparable interacting and ceaselessly moving items — a restless ever-dynamic web.

In the twentieth century, we have been able to probe ever deeper into the nature of matter. First, it was discovered that the molecules were constructed out of clusters of atoms. Then, with the help of our particle accelerators and other scientific equipment, it was found that an atom was formed by a central body, the nucleus, round which circulated one or more electrons. Finally, it was established that the nucleus itself was formed of two bodies, a proton and a neutron. The proton carried an electrical charge, and the neutron did not. Although, together, they occupy a space only about one hundred thousand times smaller than the whole atom, nevertheless they account for nearly its whole mass.

They do this, because an atom is mainly composed of empty space. It is like a vast aeroplane hanger, in the centre of which a grain of salt (the nucleus) is floating. The electrons would be represented by specks of dust whirling about the grain. The electrons are bound to the nucleus by electrical forces which try to keep them as close as possible. But nature is a never ceasing restless energy. Whenever a particle is confined, it responds by moving around. If the nucleus shortens its circular route by drawing it in more closely, it responds by concluding its journey more quickly, *plus*; and beginning on the next — five hundred, five hunded and fifty, six hundred, miles a second. It is this that makes an atom appear as a solid object.

If the aforesaid hanger were to be occupied by a propeller-driven aircraft awaiting take-off, while the pilot were away having his tea, his girl-friend might safely approach the aircraft, stretch out her arm through the space between the blades, and write with her lipstick on its nose, 'I love you.' But if she were to return later with the object of adding to her message, only to find the propeller whirling preparatory to take-off, she could not. The propeller would now not only appear to her as a solid disk, but would, in verity, possess all the characteristics of a solid disk. No way, and at no extreme of pain, could she force her arm through it. Neither can we walk through the wall of our neighbour's house, creep up behind his favourite armchair in his sitting-room, and startle him by shouting, 'E equals mc squared.' Nor stroll through a closed door, but have to go through the chore of opening it, and perambulating round its edge.

If the electrons, when confined, move at a brisk pace, what about the nucleus, confined to one far, far smaller? It outshines them by moving at a pacey forty thousand miles a second. But all this can happen only when the temperature is not too high. When the temperature increases by some hundred times, as it does in most stars, all the above structures are destroyed.

The number of electrons in an atom determine what chemical it will make. Quantum theory showed that even the particles in the atoms did not resemble solid bodies. (It was found that heat was radiated not in a continuous stream, but in a series of energy packets, each packet called a quantum, Latin plural quanta. Hence the name quantum theory.) Depending on how they are observed, they can be described both as particles and waves (a dual appearance that light often assumes). It seemed strange that something could at the same time be a particle, confined to a small space, and a wave, spread out over a large one. At the subatomic level, matter does not exist with certainty at definite places, but shows only tendencies to exist; does not occur at definite times, but shows only tendencies to occur. Particles can be waves at the same time, because they are not three-dimensional waves in the ordinary meaning of the word, but probability waves in a mathematical sense. An atomic event can never be predicted; it is possible only to predict how likely it is to happen — what are the odds. At the subatomic level, the mountains and lakes and trees and animals and buildings, including our friends and our very selves, are dissolved into wave-like patterns of proba-bilities and interconnections. Nature is a wave of inter- relations between the various parts of the whole of it. The whole of the universe is a oneness.

In their orbits, the electron waves have to be formed so that, in their confined space, they exactly make a standing wave. In music, the same phenomenon can be observed in the vibrating string of a violin or guitar or piano, or the vibrating column of air in an organ pipe or brass instrument or woodwind, of the skin of a drum. A swimmer experiences standing waves when floating in deep water by a pier or sea wall. He finds that, contrary to the expected and all the appearances, the waves are not carrying him forward at all. The water is merely lifting him up and down. It moves forward only when it encounters a sloping beach, or other similar shallows, beneath it.

Standing waves can assume only a limited number of shapes. Electron waves inside an atom can exist only in those orbits into which they fit exactly. In the simplest atom, that of hydrogen, its one electron can exist only in its first, or in its second, or in its third, etc, orbit, but not in between any of them. Usually it will be in its lowest orbit, called its ground state. From there, it can jump up into one of its higher orbits, according to how much energy it receives. It is then said to be in an 'excited state'. From this it will go back into its ground state after a while, giving off the surplus energy in the form of a photon, the parcel or quantum that carries light. And so, similarly, with the more complex atoms.

Any two atoms with the same number of electrons will be identical. They may be temporarily different if one has become excited, say by a collision with another atom. But, in the end, either will return to exactly the same ground state. Tendencies to exist, particles reaction to confinement with circular motion, atoms switching from one state to another, colliding and forming new atoms, and an essential connectedness of all phenomema — that is the whirling base upon which are founded all the things in the macroscopic, the large, world that we can see around us. It is the engine that is responsible for the numbers of atoms that bind themselves together to form molecules, from which the things we can see and hear and smell and feel are made: rocks, water, houses, vapours, trees, air, elephants, insects, men, women, and even writers on sub-atomic physics without the mathematics or the instruments, themselves to take things forward.

Such a forward move in understanding nature was made by the 'field' concept. When iron filings are sprinkled about a bar (as opposed to a horseshoe) magnet, the filings not only concentrate more thickly at the north and south poles of the magnet, but show traces of making a sweep round over a substantial area towards the poles. The field concept is a condition in the space round a charged body which will produce a force on any other charged body in that space. The same appears to be the case in the macroscopic, the large-scale, world, through the force of gravity. Instead of speaking of space being curved (both mass and energy will curve the fabric of space-time) by massive bodies, that is bodies, small or large, containing mass, we might talk of the same interdependence as is displayed in the realm of particles. Gravitational fields are set up, and felt, by all massive bodies coming within those fields. The effect is always one of attraction. This attraction not only holds us to the earth, and the earth to the sun; but also holds, though with a much smaller force, the sun to the earth, and the earth to us. Even a mouse can play its part in the gravitational game!

In the West, we have been slow in discovering this throbbing web that produces and interconnects the whole of the universe, produces all the phenomena that we perceive about us, and even the phenomenon of our ability to perceive. In Hinduism, Buddhism, and Taoism, or in their interbreeding, through the techniques of deep meditation, setting aside reason in favour of intuition, direct mystical experience, astonishing insights had been achieved many hundreds of years ago. Theirs the belief that all the things and events around them, and they themselves, were but manifestations of the same reality. This reality, the web of endless creation and destruction and recreation, was represented in countless statues of the Hindu god Shiva, his legs and arms weaving in his endless cosmic dance.

However, it takes reason to construct particle accelerators and bubble chambers and cameras, and to develop modern and future mathematics. It takes, that is, reason to explain with a greater exactitude what is going on; and mathematics to express what is going on with a greater exactitude. Mathematics! Perhaps the language of the universe. Perhaps the ultimate music to which Shiva shall dance.

Brigid and the Mountain
Sean Dorman

On my right was the mighty and bare peak of Mount Shanhoun, in shape and proportion an almost perfect pyramid. Brigid stood by the door watching me. She wore a plaid kerchief over her head and tied under her chin.

Agnes's buxom body might have become more buxom, as her mother Brigid alleged, yet secretly I was attracted by it. She was so compact, so rounded, so sturdy and vigorous, so shapely, yes, even graceful, as she moved rhythmically, as I had seen her one day weeding a field of potatoes.

Under its first title of 'Valley of Graneen', before a revision, *Brigid and the Mountain* was Recommended by the Book Society. *The Times Literary Supplement*, after a long review, summed up the book in the phrase, 'beautiful restraint'. *The Scotsman* wrote, 'His sketches are vivid and sincere. The physical aspects of the valley are described with remarkable clarity, and Mr Dorman is equally successful in his portraits of the inhabitants.' *The Sydney Morning Herald* wrote, in the course of a review of over two hundred words, 'These sketches of Donegal are delightful.' *Irish Independent*: 'Of his days in the valley, his friendships, and his talks, (Sean Dorman) has moulded a book of much charm. There is writing of grace and high degree . . . Withal, it is a notable book.' *Irish Press*: 'Sean Dorman brought with him a receptive mind, an artists's observant eye, and some writing materials. The result is . . . a very pleasant book.'

ISBN 0 9518119 8 3   Price £4.99

The Raffeen Press

Red Roses for Jenny
Sean Dorman

Red Roses for Jenny . . . What did they mean to her? A father's affection? Or a lover's desire? If they meant either to her, or both to her, then why did she throw them away? Did her mother come to hear of them? Or the wife of the man who gave them to her? And Jim, what did he think? He must have seen them, and surely he must have been disturbed. Was Jenny carrying a child, or was she not? If she were, could Jim succeed on containing the scandal and so protect his mother's feelings? Canon Moss, for all his funny ways, was wise. Was his wisdom sufficient to save them all? And, at the end of the long day, why did Jenny restore the red roses to her office desk again?

After the great success of Sean Dorman's autobiographical first novel, *Brigid and the Mountain*, initially, until revised, entitled 'Valley of Graneen' and, under that title, a Book Society Recommendation; also praised by *The Times Literary Supplement*, *The Irish Press*, *The Scotsman*, Australia's *Sydney Morning Herald*, *Irish Independent*, and many others; Mr Dorman took time off to acquire the technique of the non-autobiographical novel. The result is *Red Roses for Jenny*, with its vivid characters and driving speed of narrative. If the mountain scapes of *Brigid and the Mountain* are fine, no less fine are the seascapes of *Red Roses for Jenny*, with storm scenes as background to a love between a man and a woman no less stormy.

ISBN 0 9518119 7 5   Price £4.99

The Raffeen Press

Portrait of My Youth
Sean Dorman

Portrait of My Youth traces the earlier years of a remarkable Irish writer, Sean Dorman. The narrative, always lively, often extremely funny, sweeps the reader along on a bubbling current. There are fascinating glimpses of the British Raj in India as seen through a young child's eyes; of Algiers and Aden as seen through those of an older schoolboy; of student escapades at Oxford and in Paris in a more carefree era; of a visit to an extraordinary French family near Nice and Cannes ; of sexual shenanigans in London's bohemian Chelsea; of difficulties with an alcoholic uncle famous as an Irish playwright; of meetings with literary and theatrical notables: E. M. Forster, Granville Barker, Sean O'Casey, John Betjeman, T. S. Eliot, Barry Fitzgerald, Dame Sybil Thorndike, W. B. Yeats, Laurence Olivier, Deborah Kerr.

'Delightful, humorous, full of marvellous observation.'
Colin Wilson

At the age of fourteen, in his first term at his public school, Sean Dorman was awarded a prize as the best prose writer in the school. He was the winner of an essay competition open to the public schools of Great Britain and Ireland. After graduating at Oxford, he worked as a freelance journalist in London, contributing articles to some twenty periodicals, and ghosting six non-fiction books for a publisher. For five and a half years he edited a theatrical and art magazine in Dublin, and for twenty-six years in England a magazine for writers. His three-volume hardback, *The Selected Works of Sean Dorman*, comprises autobiography, essays, novels, short stories and verse.

ISBN 0 9518119 9 1   Price £4.99

The Raffeen Press

The Madonna
Sean Dorman

'A great twentieth-century novel.'

Judy Summers, arrested by the sound of men's voices, paused on her way to visit The Madonna. Her cheap gay cotton dress fluttered about her shapely legs. Judy Summers liked men. She liked them very much. Also, it had become imperative that she should acquire a husband . . .

They were beside the little wayside shrine. George saw that Judy's eyes were fixed on the painted Mother cradling in her arms her painted Baby. 'The birth and the feeding have been a great strain on you, darling. Don't you think that Mark ought to go on to the bottle?' Judy was shaking her head vigorously. 'I'd give my life for Mark. I feel — I feel there's something in me of The Madonna.'

George went to Rose. She drew away in hurt pride. He broke down her resistance and swept her into his arms. 'Of course you didn't mean any harm, sweetheart. I've had a very upsetting letter from Judy. I love my wife. She's the mother of my son, but it's been a great strain. You've helped me keep my sanity.' He began to rain down kisses on her brow, her cheeks, her lips. Eyes closed, she held up her face to receive them.

'*The Madonna* reads as inevitably as does Tolstoy and bears out Eliot's, "In my end is my beginning." If it reaches its proper audience, it will be read with a mixture of discovery and relief. The novel is still alive!'
George Sully

ISBN 0 9518119 6 7   Price £4.99

The Raffeen Press

Physicians, Priests & Physicists
Sean Dorman

The most potent reason for Sean Dorman's writing this book arose from the existence of his magazine *Commentary*. This monthly appeared in Dublin in the forties during five and a half years. At an average of two thousand copies a month, he felt it to be a certaintly that copies still lurked in collections both public and private, even possibly in newspapers files, there to haunt him. In his youthful pugnacity, had he somtimes overstated his case and fallen into folly? If so, the only way out was to republish his essays or editorials, with inserted toning down remarks where such seemed needed.

The essays cover the subjects of: literary censorship; cancer, heart disease and arthritis-resisting diets and exercises, including exercises underwater in a hot bath (his wife suffered from arthritis of the hip, and died of smoking and alcohol-induced cancer); the existence or non-existence of God as found in the Bible; or in the discoveries about the universe as found in the work of scientists such as Aristotle, Ptolemy, Copernicus, Galileo, Kepler, Newton, Einstein (his Special Theory, and his General Theory, of Relativity, are explained in simple terms), and the somewhat later quantum mechanics, and the twistor and superstring theories. Other essays are entitled: 'How to Rear a Baby', 'The Adventures of Marriage', Jew and Gentiles'.

ISBN 1 9012430 0 1   Price £3.99

The Raffeen Press

The Strong Man
Sean Dorman

The Strong Man, a comedy in three acts, can lay no claims either to
distinction or to having been performed on a stage. But it can claim
to have been read by a considerable number of people who have
reported that it caused them not only to smile but, on occasion, to
laugh outright. Should something that has given rise to smiles, and
even laughter, be left upon a shelf, or be entombed in a drawer? Of
course not. It should be produced in a book. Also produced in this
book are three theatre critiques. In days gone by, Ireland gave to
literature great playwrights from that seeming hotbed of dramatic
genius, Dublin University: William Congreve, George Farquhar,
Oliver Goldsmith. Since then there have been: John Millington
Synge, Samuel Beckett (both from the same university), William
Butler Yeats, Oscar Wilde, Bernard Shaw, Sean O'Casey. Well
known, but perhaps less well known than they ought to be, are Denis
Johnston and Teresa Deevy. I have devoted a critique to each of
them. Also to William Shakespeare, an Englishman, I'm told. The
trouble with William Shakespeare, is that he has been allowed,
unfortunately, to develop into a cult figure. Not only are his great
plays produced, but his lesser pieces also are reverently laid out upon
the stage, thus almost certainly denying many hours of theatre time
to others with better work to offer. Such a lesser piece, here reviewed,
is *Twelve Night*.

ISBN 0 9503455 6 3   Price £3.95

The Raffeen Press